PRINCIPLES OF BUILDING COMMISSIONING

Walter T. Grondzik
Architectural Engineer
Ball State University

WILEY

John Wiley & Sons, Inc.

Published by John Wiley & Sons, Inc., Hoboken, New Jersey
Published simultaneously in Canada

For general information about our other products and services, please contact our Customer Care Department within the United States at (800) 762-2974, outside the United States at (317) 572-3993 or fax (317) 572-4002.

Wiley also publishes its books in a variety of electronic formats. Some content that appears in print may not be available in electronic books. For more information about Wiley products, visit our web site at www.wiley.com.

Library of Congress Cataloging-in-Publication Data:

Grondzik, Walter T.
  Principles of building commissioning / Walter T. Grondzik.
     p. cm.
  Includes bibliographical references and index.
  ISBN 978-0-470-11297-7 (cloth)
  1. Building commissioning. 2. Building—Quality control. 3. Buildings—Specifications. I. Title.
   TH438.2.G76 2008
   690.028′7—dc22

                                                         2008033315

Printed in the United States of America
10 9 8 7 6 5 4 3 2 1

Contents

Contents

Preface

The seeds of this book were sown in 1992 when the Florida Design Initiative (FDI) sponsored the Second National Conference on Building Commissioning in Clearwater Beach, Florida. FDI was promoting building commissioning, among other practices, as a valuable tool that could help move buildings in the state of Florida to higher ground in terms of energy efficiency and overall performance. The purpose of the conference sponsorship was to allow Florida design professionals to meet and mingle with those from other states who were already involved with commissioning. The benefits and difficulties of commissioning were presented and discussed at the conference and a number of newly converted designers adopted commissioning as a viable means of improving building quality.

The underlying theme of the early building commissioning conferences, organized by Portland Energy Conservation, Inc. (PECI) and championed by Nancy Benner was simple—to establish building commissioning as a "business as usual" practice. In theory, when this objective was reached the rationale for the annual commissioning conferences would dissipate. The 16th National Conference on Building Commissioning was held in 2008. The need for the conferences has not evaporated, although their role has changed to one of process support versus the earlier role of philosophical conversion. At this time, building commissioning is not yet "business as usual," but neither is it an unusual business. Savvy building owners have adopted commissioning as an effective way to improve the facility acquisition process. Green building initiatives have embraced commissioning as a way of assuring quality in the delivery of high-performance buildings. Commissioning is coming of age.

ASHRAE has substantially revised its long-standing series of commissioning guidelines—with the 2005 publication of *Guideline 0: The Commissioning Process* and the just-released wholesale updating of *Guideline 1: HVAC&R Technical Requirements for the Commissioning Process*. Guideline 0 gives discipline-neutral guidance on what an effective commissioning process should look like—whether an HVAC, lighting, roof, or elevator system is involved. Guideline 1 provides the details necessary to properly implement the commissioning process with respect to HVAC&R systems. The National Institute of Building Sciences (NIBS) coordinated the development and publication (in 2006) of *NIBS Guideline 3: Exterior Enclosure Technical Requirements for the Commissioning Process*. Several additional guidelines in the NIBS "total building commissioning" series are currently in the works. Those most likely to be seen first will address fire protection and lighting systems.

Numerous organizations now embrace, support, and promote building commissioning. PECI, ASHRAE, and NIBS have been noted already, but the ranks also include the Building Commissioning Association (BCA), the National Environmental Balancing Bureau (NEEB), the California Commissioning Collaborative, and several others. Certification as a commissioning provider is available from several organizations—and in most cases the requirements for certification are anything but trivial. Commissioning is gaining traction. It may not yet be business as usual, but it is surely on the radar screens of most facility design and operations professionals.

Yet a question persists among many building designers and owners—what exactly is this thing called commissioning? Answering this fundamental question is the objective of this book. Fifteen years after the first national commissioning conference, there still seems to be a need for a consolidated source of information on the basics of building commissioning. This book attempts to be that source, and to make the fundamentals of commissioning accessible to all interested parties—to building owners and operators, to architects and engineers about to embark on commissioning efforts, and to others (such as users or suppliers) who may be called on to join the commissioning team for a particular project. This book is intended for anyone looking for "Commissioning 101." As demand for green, carbon-neutral, high-performance buildings increases, so should the numbers of people seeking to understand commissioning.

Principles of Building Commissioning attempts to clarify the underlying philosophy of commissioning: the why, what, when, and who of this process. It maps out the territory of commissioning, outlines its defining characteristics, explains its flow of processes, and demystifies its documentation. This book is very much shaped by the *ASHRAE Guideline 0* view of the world of commissioning.

Acknowledgments

I would like to gratefully acknowledge the following for their contributions to this book. Chronologically, first is Portland Energy Conservation, Inc. (PECI). Attending the PECI-organized Second National Conference on Building Commissioning opened my eyes to the irrefutable logic of commissioning as a way to improve the performance of buildings. Several months before the conference I was sent to Portland to scope out PECI and see if the Florida Design Initiative (FDI) should invest in the conference as a major sponsor. My visit was also surely a chance for PECI to scope out FDI as a potential collaborator. Luckily, all went well, and the rest is history. FDI provided major support for three of the first five national conferences. PECI is still doing outstanding work in commissioning and energy. Thanks for being an inspiration.

Second, chronologically, is the *ASHRAE Guideline 0* (and generally overlapping Guideline 1) project committee. Sitting in on committee meetings over the course of what seems like a decade (and, in fact, was close to being so) was like getting a graduate degree in commissioning. The committee members are amazingly knowledgeable and were exceptionally willing to share that knowledge—investing huge amounts of time in a totally volunteer effort. They were a fantastic group to work with, to boot. Thanks for your collective wisdom.

Third, chronologically, are the several commissioning providers who contributed freely of their time and wisdom through the interviews interspersed throughout the book. In alphabetical order these valuable contributors are:

Thomas E. Cappellin, Hanson Professional Services, Inc.
Tim Corbett, Social Security Administration
H. J. Enck, Commissioning & Green Building Solutions
Tudi Haasl, Portland Energy Conservation, Inc.
Kristin Heinemeier, Western Cooling Efficiency Center

Jere Lahey, Department of Management Services, State of Florida (interviewed by Wayne Dunn)
Jeff J. Traylor, EMCOR Government Services

Fourth, chronologically, is the John Wiley & Sons development and production team—headed by acquisitions editor Jim Harper. Thanking one's editor for patience and encouragement is a bit cliché, but in the case of this book is absolutely on target. Patience and persistence were assuredly valuable commodities. Thanks also go to copy editor Cheryl Ferguson and production editor Nancy Cintron. All have made an inherently difficult process substantially less so.

Lastly, and definitely not in chronological order, I must and do thank my family. Your support and encouragement have allowed me to essentially work way beyond the limits of what would normally be considered reasonable, without too much guilt. Carol, Michael, Tasha, Nikki, and Kelsey—thank you immensely.

Walter Grondzik, PE
Architectural Engineer
Muncie, Indiana / Tallahassee, Florida

Chapter 1

What Is Commissioning?

BUILDING COMMISSIONING

Building commissioning is simply a means of ensuring that a building owner gets the quality of facility that is expected and deserved. The word *simply* is, however, deceiving. Although the concept of commissioning is straightforward, the building commissioning process can be complex, involve numerous and continually changing players, and span the full life of the building delivery process. The purpose of this book is to describe the principles behind building commissioning and to present commissioning practices that have proven successful.

Although it may seem ridiculous to suggest that building owners are consistently not getting what they want (and often not even what they have paid for), evidence repeatedly and overwhelmingly proves otherwise. Examinations of just-occupied buildings—of all types, in many states and countries, with a variety of design/delivery systems—repeatedly reveal equipment and assemblies that are installed improperly (or were not even installed), equipment and systems that do not (and often cannot) work properly, and situations that increase energy costs, decrease building operational life, and sometimes imperil occupant health and well being.

This should not be surprising, in that the typical building is a one-off creation. Components and parts are assembled in ways that have generally been tried before, but are nonetheless specifically unique. Even if the combinations of elements are not unique, the assemblers and assembly conditions often are. That is the precise purpose of the conventional design/construction process—to meet defined needs in the context of a

1

unique site, timeline, and budget. There should be no serious expectation that a building created in this manner would work flawlessly on occupancy without some observation, testing, and tweaking of systems and assemblies. Nevertheless, owner after owner somehow expects his/her building to work well by initiating a process that does not include insightful review, testing, and adjusting. This is silly.

Extended warranties are one of the most common accessories purchased by automobile buyers. An extended warranty is essentially an insurance policy that helps ensure that the vehicle will work as expected and intended without unbudgeted and unjustified expense. This insurance is commonly and readily purchased for a mass-produced product that has undergone extensive testing and quality control procedures—and one that involves a fraction of the investment in a typical building.

Owners should view commissioning as the conceptual equivalent of an extended warranty for their buildings—remembering that these are buildings that are not mass-produced and often have seen little in the way of formal quality control. Carrying this analogy a step further, ongoing building commissioning is the equivalent of scheduled maintenance for a car; a way of avoiding unpleasant surprises that deprive an owner of effective use of his/her car (building) and/or require unplanned emergency repair and remediation costs.

The idea of the navy commissioning its ships is often used an example of building commissioning. This is partially true, with sea trials (the validation process) preceding "commissioning" (formal acceptance into the fleet). Quality control for shipbuilding is typically independent of (although related to) the performance validation activity. The movie image of ship commissioning is also more dramatic—"I don't care what the pressure gages say, Mr. Roberts, give me 35 knots now!"—than the typical building commissioning experience. Yet the idea is precisely the same; it makes infinitely more sense to detect problems or failure in a trial run conducted on your own terms than in a crisis (a battle, in the case of the ship, or a building occupied by 1,200 highly paid and previously productive professionals).

One way to view commissioning is to consider it as a partial step toward integrated practice. In an integrated practice, disciplinary boundaries and walls around project phases are broken down such that all participants are working seamlessly toward a common goal, without the communication gaps and suspicions that can arise from the conventional design-bid-build approach (Elvin 2007). The commissioning process, and more specifically, the commissioning team, can act as an effective project integrator during the transition to fully integrated practices.

THE BUILDING ACQUISITION PROCESS

As will be described in the next chapter, commissioning is a process that parallels and integrates with the conventional design-construct-occupy process for buildings. Although there are several important and common variations of this process, the conventional design-bid-build approach will be the basis for discussion in this book. The principal phases in this process are shown in Figure 1.1. These are the building acquisition phases identified by *ASHRAE Guideline 0: The Commissioning Process* (ASHRAE 2005).

The acquisition process for a new building consists of a sequenced series of activities intended to provide a facility that meets the owner's needs. During the *predesign phase* of the process, these needs are identified, honed, and documented. Historically, the result of this phase is an owner's program (or brief) that becomes the foundation for design efforts. Such a program may be developed solely by the owner, by the owner in cooperation with a programming specialist, or by the owner in conjunction with the design team. An incomplete or inaccurate program will lead to an incomplete or partially functional building. Incomplete programs may

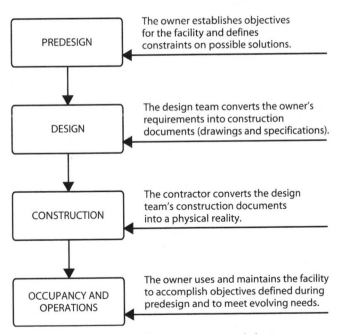

Figure 1.1   Phases in the conventional design-construct-occupy sequence for a building project.

involve either missing spaces (too few offices or conference rooms) or incompletely defined spaces (classrooms without audiovisual capabilities, with no flexibility for evolving functions, and the like). Serious ambiguity regarding the intended quality of a space or facility is common in many owners' programs.

During the *design phase*, the design team (architects, engineers, and often specialists) attempts to convert the owner's program into plans and specifications (construction documents) for a facility that will reflect the needs and desires outlined in the program. Budget and schedule are often overriding constraints. The extent of communication between the owner and the design team during the design phase can vary greatly from project to project. The intent of this phase is to prepare contractually binding documents that can be successfully used to convert an idea into a physical reality. During the course of design development, thousands of decisions will be made based on hundreds of assumptions, calculations, and precedents. The design team's values and desires will be superimposed on those of the owner. Decisions made during design will affect both the constructability and operability of a facility.

During the *construction phase*, a contractor attempts to convert the construction documents into a physical entity. Although the contractor usually contracts with the owner, the contract is to execute the design team's documents. The owner is free to ask that changes to the drawings and specifications be made to accommodate second thoughts or evolving needs—but such changes typically come at a substantial cost in time and money. The contractor's values and desires will be superimposed on those of the owner (as filtered through the design team). It is not unusual to see adversarial relationships between the design team and the contractor creep into the construction phase, often as a result of varying interpretations of the construction documents and differing opinions regarding the expected quality of materials and workmanship. Communications between the contractor and design team can vary widely from project to project, as can the extent of observation of the construction process by the designer.

Occupancy and operations is an extensive phase wherein the facility is complete and is used, ideally, as originally envisioned in the owner's program. The value and utility of the facility is maintained or enhanced through owner decisions regarding maintenance, operations, and remodeling. Decisions made during the design and construction process can dramatically affect the maintainability and usability of a facility, although such implications are often not obvious to the owner during the predesign phase or made clear to the owner during the design phase. Substantially more investment may be required to operate and maintain a facility across its life span than was required to obtain the facility. It is

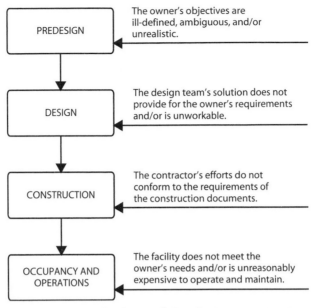

Figure 1.2   Each phase of the design-construct-occupy process can lead to problems for a building owner.

not unusual to find less guidance given a building owner about operations and maintenance than is provided to the typical new car buyer.

Figure 1.2 summarizes some of the many glitches that can—and often do—occur within the individual phases of the design-construct-occupy process. Although it is useful to identify these separate phases, the complete building acquisition process involves all the phases linked into a continuous (or nearly so) sequence. Glitches can also occur in the transition from one phase to another, as suggested in Figure 1.3. A well-developed owner's program may not be clearly transmitted to the design team—or key members of the team (e.g., consulting engineers) may never see the program. The design team's solution may be inadequately conveyed to the contractor (through poor drawings, ambiguous specifications, or last-minute changes in subcontractors). It is common that the handoff of building from contractor to owner is done with very little usable supporting information—for example, where do I get replacement lamps, and how do I reach the fixtures?

The facility acquisition process is essentially a relay of information from the owner back to the owner—with many intervening parties and contracts, and usually substantial intervening time. There is simply too

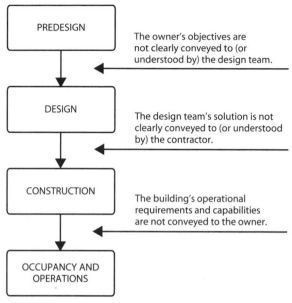

Figure 1.3   Poor coordination between phases of the design-construct-occupy process can lead to problems for a building owner.

much that can go wrong in the process to trust to luck. Experience also shows that too much can go wrong to trust design team and/or contractor assurances that commissioning is "already being done, but under a different name," or that "commissioning will just cost you money and slow things down." The sidebar provides examples of some of the things that have been found wrong in reasonably simple buildings that were not commissioned. Such situations are legendary.

Who Needs Commissioning?

One of the most telling arguments for building commissioning came out of a Florida Solar Energy Center study of uncontrolled airflow in buildings (Cummings et al. 1996). A large number of small commercial buildings (medical office, strip mall, etc.) were carefully investigated to see how critical an issue unintended air flow (leakage) might be in such buildings. The results were amazing. Of the 70 buildings investigated, 69 were found to have some defect or flaw that would permit uncontrolled airflow.

These were not just subtle flaws detectable only by trained scientists. Included were problems such as toilet exhaust fans dumping air into return air plenums, no ceiling/roof insulation (where there should have been), vapor retarders that did not continue above suspended ceilings, fans running backward, and gaps between sections of ductwork. Interestingly, none of the problems seemed to be obvious enough to raise preinvestigation concerns on the part of owners/occupants. Most of the problems, however, increased energy usage and/or decreased comfort day after day after day. Many of the problems would most likely lead to indoor air quality problems such as mold/mildew growth.

Two telling extracts from the report suggest the nature of problems commonly encountered in real-life buildings:

> Consider two examples of duct leaks that were occurring in the outdoor air ductwork in one recently tested commercial building. In one case, an outdoor air duct on the first floor stopped short of the grille at the exterior wall of the building leaving a 2 inch gap. As a consequence, about 75% of the "outdoor air" was actually being pulled from the building. In the other case, outdoor air ducts went from two second floor air handlers to panels in exterior walls where there were supposed to be exterior outdoor air grilles. However, there were no grilles—just a solid brick wall!

> In one case, a government office ... located next to a cocktail lounge experienced air quality problems, including tobacco smoke and smells of cleaners resulting from air transport between units. Above the suspended t-bar ceilings, the ceiling spaces are well connected to each other by openings in the fire wall that totaled over 30 square feet in size. In addition, the bathroom exhaust fans from the bar discharged into the ceiling space of the office. Once in the ceiling space, air contaminants were transported into the office space by means of large return leaks.

Such findings of defects in existing (and often relatively new) buildings have been replicated in study after study. Who needs building commissioning? Most owners need building commissioning.

WHAT BUILDING COMMISSIONING IS AND IS NOT

Building commissioning is essentially a communication and validation process that begins as early in the building acquisition process as possible and continues through owner occupancy. There is no magic involved with commissioning. Communication is the key to success on any multiphase project involving literally hundreds of decision makers and spanning several months to several years. The commissioning process facilitates well-documented communication among owner, designer, contractor, and operator. This may seem trivial, but its value is inestimable.

Validation is the second key element of the commissioning process. Rather than assuming that things are as they should be, selected decisions, assemblies, equipment, and operations are tested (conceptually and/or physically) to ensure that they meet the needs and expectations of the owner.

Obviously, commissioning involves more than these key aspects, and the details will be presented in subsequent chapters. In a nutshell, however, commissioning consists of communicating requirements and expectations and verifying that they are met by the various parties to the building acquisition process.

Commissioning is not an additional phase of a project. Commissioning is not an isolated testing event. Commissioning is not TAB (testing, adjusting, balancing). Commissioning is not equipment start-up. Commissioning will likely involve TAB, equipment start-up, and testing of various types, but these are just a part of the larger whole of the commissioning process as it occurs throughout all phases of a project.

WHAT BUILDING COMMISSIONING CAN DO

Properly executed, building commissioning begun during predesign should consistently be able to do the following:

- Help clearly define an owner's project needs and expectations.
- Validate design solutions against defined owner requirements.
- Reduce change orders brought about by poor communication.
- Verify (before occupancy) that key building components work as intended.
- Provide appropriate training for building operating personnel.
- Provide useful project documentation at owner takeover.
- Provide benchmarking of building performance for ongoing effective and efficient operation and maintenance.

WHAT BUILDING COMMISSIONING CANNOT DO

Building commissioning is not a replacement for the conventional building acquisition process—it is a supplement to that process. As a result, commissioning should not be expected to do the following:

- Make up for ill-conceived project objectives and expectations.
- Make up for an inadequate design and/or construction budget.

- Design a building or systems in lieu of the design professionals of record.
- Construct a building in lieu of the contractors of record.
- Repair major problems and deficiencies with systems or equipment.
- Operate or maintain a building.
- Correct project-long problems through last-minute intervention at the end of construction.

TOTAL BUILDING COMMISSIONING

Building commissioning has generally been viewed (at least historically) as being applied primarily to mechanical systems (especially HVAC) and perhaps electrical systems. This mental bounding of the scope of commissioning to dynamic systems has limited its potential usefulness in providing verified quality for building systems and assemblies. The idea of *total building commissioning* was developed to overcome this limited view and expand the value of the commissioning process. Total building commissioning does not require or suggest that every element and system in a building is to be commissioned. Rather, it attempts to provide procedures and tools by which nonelectrical/mechanical systems (such as walls, roofs, furnishings) may be commissioned—if so desired by an owner.

The National Institute of Building Sciences (NIBS) is spearheading an ongoing effort to develop guidelines for the commissioning of a range of building systems—including telecommunications, lighting, plumbing, envelope, interiors, and fire-protection systems. Specific guidelines will be developed by cooperating professional associations (such as the National Fire Protection Association and the Building Enclosure Technology and Environment Council). In general, the commissioning *process* (as described herein) will be the same for any system or assembly to be commissioned. Only the details of implementation will vary. The first of the NIBS-developed commissioning guidelines deals with building envelopes (NIBS 2006).

The advent of total building commissioning has generated some debate as to whether a static assembly or component can be "commissioned." This book views commissioning as a quality-control verification and validation process; thus, there is no reason why a generally static element—such as a wall, roof, or partition—cannot be commissioned. Some elements may require great ingenuity in the development of appropriate "tests," but virtually every element has some set of parameters that are often worthy of validation during design and verification following installation. This is the fundamental intent of building commissioning.

INTERVIEW

Views on the Commissioning Process

Tudi Haasl, Senior Technical Advisor, Portland Energy Conservation, Inc., Portland, Oregon

Q. As a long-term promoter of building commissioning, how would you describe the landscape for commissioning services today (as opposed to, say, 10 years ago)?

A. *The landscape has changed dramatically. There is much more demand for both new and existing building commissioning. This is due, in part, to building owners' desire to be "greener." The U.S. Green Building Council's Leadership in Energy and Environmental Design (LEED) Green Building Rating System™ includes some requirements for commissioning, and this has driven the market to some extent. Some states are requiring commissioning for their public buildings in an effort to meet executive mandates. Market transformation is happening at a rapid rate due to high energy prices and concerns about global warming; thus, commissioning and retrocommissioning are seen as a strategy to help reduce both expenses and environmental degradation. A number of utilities across the country provide incentives to building owners for existing building commissioning—10 years ago, this wasn't in the mix. There is more and more activity and interest in the private sector, especially from large commercial building owners and "big box" retailers.*

The National Conference on Building Commissioning is now a venue where all interested parties come together to learn and participate (it's no longer just a commissioning provider conference).

Q. What is the most positive thing about today's commissioning environment?

A. *A recognition on all levels—from engineers and building owners to maintenance service providers and building operators—that there is great value in commissioning, leading to opportunities to obtain and provide better service for the built environment. There is a trickle-down effect happening to some degree. Maintenance service contractors are looking for ways to improve their offerings; they see value in doing building tune-up activities (targeted or focused commissioning) that lead to energy cost savings for their customers. They are beginning to include operational checks in their contracts, as well as the usual maintenance checks.*

Q. What is the most negative thing about today's commissioning environment?

A. *The greatest concern in the midst of all this commissioning demand is getting well-trained, ethical commissioning providers into the field so commissioning doesn't become watered down or be seen as a barrier to find a way around. Lack of a solid infrastructure has always been a* concern, but is now more important than ever. Also, several entities are now providing certifications with different levels of rigor. This is confusing to those market players who are in the position of purchasing services.

Q. What seems to be the biggest misperception regarding commissioning among owners?

A. Some of the same misconceptions exist today that have existed since the beginning. It's too expensive; I thought I was already paying for this; It's just another layer that will cost me more money, etc.

Q. What seems to be the biggest misperception regarding commissioning among design professionals?

A. Excellent, high-end design firms understand the importance of testing their designs and are usually good at commissioning. They do it themselves and see it as a part of the whole effort and a service to the customer. However, in this quick-turn-around and low-budget environment, commissioning can be seen as a barrier for designers. We are still up against difficulties in getting commissioning in during the design phase of new construction. Even LEED for New Construction has fragmented the process (the very process that, by definition, is a process of integration) by making "Design Phase Commissioning" a choice (extra point) instead of making it a requirement.

Q. What seems to be the biggest misperception regarding commissioning among contractors?

A. It's going to hold things up and cost us a lot of money.

Q. What do you recommend as a means of breaking the ice on commissioning for the novice owner?

A. Establish a relationship with a trusted provider—get recommendations from peers. Know what economic resources are out there, such as utility programs. Do your homework. Understand what the commissioning deliverables should be and look like—several are listed on the California Commissioning Collaborative Web site (www.cacx.org) for both new and existing building commissioning. There is a new guide available for building owners interested in retrocommissioning that was funded by EPA, called "A Retrocommissioning Guide for Building Owners." Check www.peci.org/CxTechnical/resources.html for the top five resources (the EPA-funded guide is among them). Stay involved and in the driver's seat.

Understand what strategies will be put in place so that the benefits from the commissioning process last. Ultimately, it's about getting persistence.

Q. What developments do you see on the horizon for building commissioning over the next 5 to 10 years?

A. There should be continued growth in the industry. Commissioning is still not business as usual, but is starting to move in that direction. Maybe in another 10 years it will be. Based on current trends, commissioning *for existing buildings will continue to be a strong focus over the next 5 years as we all struggle with global warming and high energy prices. Persistence tools and strategies will be more in demand, since commissioning benefits may degrade over time, especially without adequate training and documentation. There will be higher interest in and hopefully demand for monitoring- or persistence-based commissioning services. More protocols will be developed for testing low-energy building equipment and installations.*

Q. Do you have any words of wisdom for those newly embarking on the building commissioning process? Or for seasoned professionals?

A. Know what utility programs are out there so you can leverage them. Get trained in controls and integration issues. Become skilled in doing design reviews and hands-on-testing of equipment and systems. Get tooled up, and have a good set of portable data loggers in your tool kit.

Most important is one's EQ, or emotional intelligence quotient. IQ means you can take tests and do well in school in all subjects. EQ is what you need in order to deal with people, resolve conflicts, and keep your ego at bay. It is often overlooked, but positive interpersonal and group interactions are a big piece of a successful commissioning project.

Q. What are some of the greatest challenges as the commissioning industry moves forward?

A. The following are important:

- *Developing a highly trained commissioning provider work force*
- *Continuing to educate owners so that owners will educate each other*
- *Standardizing the commissioning process without jeopardizing creativity and ingenuity*
- *Working more closely with controls contractors to deliver what their products are capable of doing and to increase the quality of documentation for the operating sequences*
- *Getting persistence strategies and tools to be a first thought, not an afterthought*
- *Obtaining more funding for research to measure and verify the benefits of high-quality commissioning efforts*
- *Revising LEED to require, at a minimum, an integrated commissioning process for the building energy-using systems*

REFERENCES

ASHRAE. 2005. *ASHRAE Guideline 0: The Commissioning Process*. The American Society of Heating, Refrigerating and Air-Conditioning Engineers, Atlanta, GA.

Cummings, J. et al. 1996. "Final Report: Uncontrolled Air Flow in Non-Residential Buildings" (FSEC-CR-878-96). Florida Solar Energy Center, Cocoa, FL.

Elvin, G. 2007. *Integrated Practice in Architecture*. John Wiley & Sons, Hoboken, NJ.

NIBS. 2006. *NIBS Guideline 3: Exterior Enclosure Technical Requirements for the Commissioning Process*. The National Institute of Building Sciences, Washington, DC. www.wbdg.org/ccb/NIBS/nibs_gl3.pdf.

Chapter 2

The Commissioning Process

COMMISSIONING IS A PROCESS

The concept of building commissioning is simple: Undertake a series of interrelated actions to ensure that a building produced through the design-construct-occupy process actually meets the owner's requirements. These actions typically involve a diverse range of services to minimize unpleasant surprises when the owner assumes control of and begins to operate a facility. Commissioning is not, and cannot possibly be, a week-long special event. Commissioning is not an additional project phase. Commissioning is an ongoing process—a process that parallels and complements the entire building acquisition process. This chapter outlines the commissioning process and is organized around the phases of the building acquisition sequence identified in Chapter 1. Subsequent chapters will focus on specific activities that make up the commissioning process.

There is no single "right way" to undertake commissioning. The commissioning process should shift, adjust, and adapt to each project's needs and context. Nevertheless, there are certain attributes that clearly define commissioning and set it apart from other quality-control procedures, professional services, and design/construction activities. This chapter attempts to identify these key attributes—elements that should be included in commissioning for any project. This view of the commissioning process is in accord with *ASHRAE Guideline 0: The Commissioning Process* (ASHRAE 2005). Customizing the details of the commissioning process to best fit a specific project is the responsibility of the owner and the commissioning authority.

Ideally, the commissioning process begins early in the predesign phase. The reason why this is so important should be made clear in the following discussion. This is not to say that commissioning cannot ever start during the design phase (or, much less desirably, during construction). Beginning the commissioning process after predesign, however, introduces serious impediments to success that must be overcome and may be very difficult and expensive to counteract. Starting later and later shifts more and more participants into a defensive or adversarial position. The design team and/or contractor may only see unanticipated changes to their normal, expected, and already-under-contract tasks—rather than the cooperative (and contracted for) roles envisioned by the commissioning process.

PREDESIGN PHASE

Commissioning activities during the predesign phase set the stage for the commissioning process and are critical to the success of commissioning. The two primary activities are development of the Owner's Project Requirements and preparation of a draft Commissioning Plan. *(Note: capitalized terms in the text are used to indicate a formal document or commissioning process deliverable.)* Although the Owner's Project Requirements will likely be refined in subsequent project phases, this document is essentially completed during predesign. The Commissioning Plan is begun during predesign, but expands substantially (and is amended as necessary) during subsequent project phases. A "commissioning authority" is involved in developing both of these documents.

The commissioning authority is the coordinator of the commissioning process. Bringing a qualified commissioning authority on board is the first step an owner should take after deciding to commission a project. The commissioning authority will assist the owner in defining a scope and budget for commissioning (a key part of the Commissioning Plan) and in developing appropriate project requirements. Further information on the commissioning authority is provided in the following chapter.

Owner's Project Requirements

The Owner's Project Requirements (OPR) is a formal document prepared by the owner (or someone designated by the owner) that captures the needs and expectations for a proposed facility. The OPR is the basis upon which all validation activities related to commissioning are based. A well-developed and complete Owner's Project Requirements document is an absolutely critical first step in the commissioning process.

The owner communicates his/her expectations for a facility to the design team through the Owner's Project Requirements. In later project phases, the commissioning team validates decisions, equipment, assemblies, and systems with reference to this document. The OPR should clearly define and describe—without getting into design methods or solutions—those aspects of quantity and quality that will make a proposed facility a success. A good Owner's Project Requirements document will capture all of this information in a form that can be readily disseminated to the design team, commissioning team, and contractor (in the case of the contractor, for informational, not contractual, purposes).

The term *Owner's Project Requirements* is relatively new. It was coined to clearly distinguish this document from the more familiar owner's "program" (or project "brief"). The content historically developed for an owner's program provides the foundation for the Owner's Project Requirements. The typical owner's program, however, is too ambiguous and incomplete to serve as a useful reference for commissioning. The expected quality of a proposed facility and its constituent parts is usually poorly described in owner's programs. Owner's programs seldom discuss owner capabilities and expectations for the operation of a facility; they rarely deal with reliability, maintainability, replacement, and operations and maintenance staff capabilities. Further, owner's programs often address perceived solutions instead of functional requirements. Excerpts from an actual owner's program are shown in the following highlight to illustrate these points.

Why a Conventional "Program" Is Not Sufficient

The materials presented here are an excerpt from an actual owner's program for renovation and expansion of a university classroom/office building. Although only a small part of a much larger program, the excerpt is interesting in that it represents *all* that the program had to say about this particular space (its quantity and quality aspects). Any quality aspects above and beyond what appears here would be included only because they were a part of the design team's normal approach to design or represented current normal practice. In most walks of life, this would be called getting something "by accident."

Upper Division Studios

Number of stations = 32
NSF per station = 70
NSF per space = 2,240
Number of spaces = 5
Total NSF = 11,200

Use description: CAD, mechanical and freehand drawings, meetings, model construction, group and individual instruction, and
critique

Furniture and equipment: extremely durable and lightweight
drawing and layoff tables (two options noted); 32 adjustable
cushioned stools with back rest; 32 secure storage bins for
drawings and equipment

Criteria relationships to other spaces: student commons/lounge;
other studios; primary horizontal and vertical circulation

Special requirements: power and data requirements for complete
CAD studio environment capability; continuous power/data
molding placed at approximately 24–30 inches AFF (current
ones are too high); lighting capability to accommodate both
paper and monitor use; direct physical connection to adjacent
jury and computer spaces in suite; as much shaded daylight
as feasibly possible; primary pin-up wall space would occur in
adjacent jury room

From a design and commissioning perspective, this list is very
sketchy, and includes solutions as well as some desired outcomes.
The inclusion of solutions tends to tie the design team's hands
(making alternative solutions less likely to be considered). Worse
yet, many important aspects of the proposed space are not addressed
at all: acoustics, thermal comfort, indoor air quality, flexibility,
and maintainability. Is it really OK to fall back to "defaults" on
these concerns, or were they simply overlooked? The purpose of the
Owner's Project Requirements document is to tie down as much of
the owner's thinking about a proposed building as possible—with
input from designers, contractors, and users. Developing such a
useful starting point for design efforts might be justification enough
for undertaking the commissioning process.

Commissioning Plan

The Commissioning Plan is a formal document prepared by the commissioning authority (with substantial input from the owner and other
commissioning team members) that outlines and defines the commissioning process envisioned (and eventually implemented) for a specific
project. During predesign, the Commissioning Plan will primarily address
the proposed scope and budget for the commissioning process. It is

critical, however, that commissioning activities to be conducted by design professionals during the design phase be defined in sufficient detail that they may be included in professional services fee proposals and agreements. Requirements for participation of owner representatives as part of the commissioning team should also be described.

The scope and budget for commissioning must be tied down during predesign. In terms of scope, a sense of what systems and what major elements of those systems are to be commissioned is needed. For example, will commissioning address only the HVAC&R system—or the HVAC, electrical, security, fire protection, and roofing systems? If the HVAC&R system is to be commissioned, will this include heating and cooling sources, air distribution, terminal units, controls? It will not be possible to be very specific about equipment and systems at this stage, as design decisions have yet to be made. Nevertheless, a good idea of what commissioning will encompass can (and must) be established.

A reasonable approach to establishing the scope of commissioning for a project is to combine the advice of the commissioning authority with the experiences of the owner. It makes sense to commission whatever has proven to be a headache in the past, whatever is crucial to the mission of a facility, whatever is critical to life safety and health, and elements that can cause unnecessary and hard-to-predict expenses for energy or repairs. An external driver for commissioning—such as the U.S. Green Building Council's LEED (Leadership in Energy and Environmental Design) green building certification program—may help shape the scope of commissioning efforts. When establishing the scope and objectives of commissioning, remember that an owner does not require a particular fan or a specific pump; thermal comfort or good indoor air quality is the objective. A focus on commissioning (verification) of outcomes is preferred over a focus simply on commissioning of equipment and assemblies.

The commissioning process budget will be established with input from the commissioning authority based on a defined scope of commissioning. The budget should be realistic and be firmly allocated. A decision to use the commissioning process must be coupled with a determination to fund the process. Although the costs of commissioning have been reported to cover a wide range (including negative net costs by some reliable sources), it is fair to say that there is no such thing as a free lunch. Commissioning activities must be funded, and this funding allocation cannot be seen as a convenient pocket to dip into for contingencies or to make up for cost overruns elsewhere in a project. Some interesting findings regarding commissioning costs and benefits are presented to encourage appropriate funding of this useful process shown in the previous highlight titled Why a Conventional "Program" Is Not Sufficient.

Commissioning Costs and Benefits

The following information is from a summary study by Evan Mills, Lawrence Berkeley National Laboratory (Mills 2004):

> We develop a detailed and uniform methodology for characterizing, analyzing, and synthesizing the results. For existing buildings, we found median commissioning costs of $0.27/ft^2, whole-building energy savings of 15 percent, and payback times of 0.7 years. For new construction, median commissioning costs were $1.00/ft^2 (0.6 percent of total construction costs), yielding a median payback time of 4.8 years (excluding quantified nonenergy impacts).

> These results are conservative insofar as the scope of commissioning rarely spans all fuels and building systems in which savings may be found, not all recommendations are implemented, and significant first-cost and ongoing nonenergy benefits are rarely quantified. Examples of the latter include reduced change-orders thanks to early detection of problems during design and construction, rather than after the fact, or correcting causes of premature equipment breakdown. Median one-time nonenergy benefits were −$0.18/ft^2-year for existing buildings (10 cases) and −$1.24/ft^2-year for new construction (22 cases)—comparable to the entire cost of commissioning.

> Deeper analysis of the results shows cost-effective outcomes for existing buildings and new construction alike, across a range of building types, sizes and pre-commissioning energy intensities. The most cost-effective results occurred among energy-intensive facilities such as hospitals and laboratories. Less cost-effective results are most frequent in smaller buildings. Energy savings tend to rise with increasing comprehensiveness of commissioning.

DESIGN PHASE

The key commissioning process activities during the design phase include development of the Basis of Design, continued development and expansion of the Commissioning Plan, inclusion of commissioning activities in the Construction Documents (primarily the specifications), and verification of design documents. The primary objectives of commissioning during the design phase are to verify that design decisions support fulfillment of the Owner's Project Requirements and to ensure that required commissioning activities are properly addressed and clearly conveyed to the contractor.

Basis of Design

The Basis of Design is a formal document that is prepared by the design team, reviewed by the commissioning team, and accepted by the owner.

The purpose of the Basis of Design is to capture the underlying thinking about systems design that leads to the Construction Documents that are eventually provided to the contractor. The Construction Documents (drawings and specifications) show what the contractor is to do, but not why decisions leading to these documents were made. The Basis of Design captures this *why* aspect of the design process, which can be an exceptionally useful resource for the commissioning process and a valuable tool for ongoing building operations and maintenance.

The extent to which the hundreds upon hundreds of decisions made during the design process are recorded will vary from design firm to design firm. It is currently rare, however, for a design firm to record decisions in a format suitable for outside use. The Basis of Design is intended to be seen and used by various members of the commissioning team. It should be legible, be well-organized, include key assumptions made during design, note the calculation methods used and capture input to such calculations, indicate equipment types and manufacturers used as the basis for design, and provide a narrative description of systems and their intended operation. The narrative descriptions will evolve in complexity as the design process moves along.

Too many decisions are made during the design process to permit any rational comparison of output (the Construction Documents) to input (the Owner's Project Requirements and design team values) without knowledge of the intervening steps. The Basis of Design provides this connecting knowledge. The Basis of Design allows the commissioning team to validate decisions (system type, equipment type, capacity, maintenance needs, etc.) against the owner's expressed needs and expectations by establishing a cause-and-effect information trail. The Basis of Design will be used to verify the Construction Documents against the Owner's Project Requirements during the design phase. During the construction phase, the Basis of Design can be helpful in determining the validity and consequences of contractor-proposed substitutions and owner-proposed changes.

It may sound as if the Basis of Design and/or Owner's Project Requirements act to handcuff design and construction decision making. This is true to the extent that adherence to these documents can prevent unwise decisions from being made during design, construction, or operation of a facility. Neither document is, however, immutable. Both should (and will) be amended as necessary to reflect changes in the owner's philosophy, budget, schedule, or needs. Neither should be changed, however, without thinking through the implications of suggested changes. The collective wisdom of the commissioning team acts to maintain the integrity and ensure the usefulness of these benchmarking documents.

Commissioning Plan

The initial Commissioning Plan, developed during predesign, is substantially expanded during the design phase to thoroughly address commissioning activities to be completed during the construction phase. Commissioning activities that will occur during the owner occupancy and operations phase are outlined at this time.

Construction Documents

For the contractor to willingly participate in the commissioning process— which is absolutely essential—the contractor's roles and responsibilities must be spelled out in the Construction Documents. The drawings and specifications that constitute the Construction Documents generally define what is "in contract" (and, by exclusion, what is not). All relevant commissioning activities that involve the contractor must be placed "in" contract and the contractor must be given a fair opportunity to include these activities in the bid proposal. Commissioning activities requested of the contractor after the fact (after contract signing) will typically be treated as change orders, which will negatively influence the project budget and probably the project schedule.

The commissioning requirements detailed in the drawings and specifications should be adequate to permit the contractor to include them in the construction budget and schedule and in any necessary subcontracts. Commissioning process meetings, various contractor-performed tests, contractor documentation requirements, requirements for manufacturer's training or performance certifications, providing access and coordination for cross-system testing, requirements for third-party testing, and the like must be spelled out in the Construction Documents. It is fairly common for such contractual requirements to be stated as allowances to accommodate uncertainty and provide flexibility.

Several conflicting perspectives on testing and verification are held by commissioning authorities. Some recommend the testing of all critical elements. Others strongly suggest that verification and testing be based on a scientific sampling approach. For example, 10 percent of lighting dimmer controls might be randomly tested; if more than 10 percent of the dimmers that are tested fail, then an additional 20 percent of the dimmers would be tested. Whatever approach to testing is to be used on a particular project should be clearly spelled out for the contractor. What will be done with items that fail testing and verification must also be spelled out, including a clear statement of responsibility for correction of defects and for retesting expenses.

The issue of what to do with elements that fail verification tests leads to a further question. To what standard should items be tested? Contractually, the contractor can only reasonably be held to compliance with the Contract Documents. Historically, these documents say little about complex performance issues. The Construction Documents typically say nothing (directly) about many of the owner's requirements and expectations for a project. It is normally assumed that the design team has successfully linked the Contract Documents the team has developed to the project requirements developed by the owner. This is not always the case. It is wise to have a contingency plan in place to address situations where verification indicates that a particular element or system meets the requirements of the Contract Documents but does not meet the Owner's Project Requirements.

Verification review of the Basis of Design and Construction Documents is a recommended design-phase commissioning activity. Such a review (generally based on statistical sampling) does not infringe on the design team's legal and ethical responsibility for the accuracy and completeness of the Construction Documents. The design team is the party of record. Such verification does, however, provide for an overview of design decisions using the Owner's Project Requirements as a benchmark and takes advantage of the diverse eyes of the commissioning team members. The owner is the final arbiter in situations where the design team and the commissioning authority hold irreconcilable differences of opinion as to design adequacy.

Verification review of Construction Documents will typically address the ability of selected portions of the design (as it evolves) to meet the Owner's Project Requirements. Full document review is unreasonable, uneconomical, unnecessary, and practically impossible. The elements to be reviewed may be selected by the commissioning authority because of their importance to critical project requirements and/or by a statistical sampling process (similar to that described above for equipment/systems testing and verification).

CONSTRUCTION PHASE

The primary commissioning process activities during construction include verification testing of selected equipment and systems, training of the owner's operating and maintenance personnel, preparation of a Systems Manual, updating of the Commissioning Plan, and updating of the Owner's Project Requirements. Another commissioning process activity that should be undertaken during the construction phase is commissioning team assistance to minimize the scope of the contractor's punch list—the list of necessary corrections or fixes developed by the owner near the point of project turnover. This particular activity involves tapping into the commissioning team's expertise and communications skills.

Equipment and Systems Verification

Although many owners and designers have in the past viewed commissioning as simply systems and equipment testing (and some diehards still do), this important aspect of the commissioning process is really only one part of an ongoing series of verifications. The importance of systems and equipment verification will vary from system to system, with mechanical and electrical systems typically reaping the most benefit and demanding the most extensive verification activities.

Several distinct levels of equipment verification should be part of the construction-phase commissioning process:

- Verify that the correct equipment has been delivered to the project.
- Verify that such equipment has been properly installed.
- Verify that equipment operates properly within its own context.
- Verify that equipment operates properly within the context of its larger system.
- Verify that equipment operates properly within the context of an interdisciplinary system.

These verifications should be conducted sequentially, as it makes no sense to attempt to verify that the wrong piece of equipment running backward is doing what it is supposed to be doing with respect to indoor air quality. Equipment and systems verifications are generally conducted using checklists developed specifically for the given equipment/system and customized to suit a particular project. Verification tests are typically conducted by the contractor under the watchful eye of the commissioning authority and/or other members of the commissioning team.

Training

Training of owner's operations and maintenance personnel is a critical aspect of construction phase commissioning. The requirement for and scope of such training is specified in the Construction Documents. Training is conducted by the contractor, manufacturer's representatives, or other specialists as defined in a formal Training Plan that becomes a part of the overall Commissioning Plan. The commissioning authority will verify that appropriate training has been provided and will often assess its effectiveness.

Systems Manual

Systems Manual is the preferred term for what has historically been called the O&M Manual (Operations and Maintenance Manual). The purpose of the Systems Manual is to provide the owner with a well-structured, easily accessible, and useful source of reference materials dealing with the various building systems—and a document that is actually organized around such systems. The Systems Manual should contain everything necessary for the owner's staff to intelligently run and maintain the building systems, including the following:

- Owner's Project Requirements related to each system
- Basis of Design information for each system
- Control narratives (where appropriate)
- Completed verification checklists for system and system components
- Operations information (procedures, benchmarks, cautions)
- Maintenance information (procedures, intervals, parts)
- Training materials (for new hires and refreshing of skills)

Commissioning Plan Updates

During the construction phase, the Commissioning Plan will be updated to reflect any necessary changes brought about by approved substitutions, change orders, or equipment/system testing failures (those failures that are specifically accepted by the owner). The Commissioning Plan is also updated to more fully address commissioning process activities that will occur during the occupancy and operations phase. Updates are also made to the Owner's Project Requirements and Basis of Design—as needed to reflect construction phase decisions.

OCCUPANCY AND OPERATIONS PHASE

During the occupancy and operations phase, an owner assumes responsibility for the use of his or her building. Although there is a powerful conceptual break at this point—from acquiring a facility to using it—this is not a good point to break off the commissioning process. Most experts recommend that commissioning activities continue into this usage phase at least until all warranties have expired—and often much longer, essentially transitioning into an "ongoing commissioning" process with activities that will occur periodically during the life of the building.

Typical commissioning process activities during the occupancy and operations phase include the following:

- Completion of any deferred systems/equipment verifications
- Completion of any deferred training
- Providing assistance to the owner for any warranty-related problems
- Benchmarking in-use building performance for future use by the owner

Details on commissioning process documents and verifications are presented in the chapters that follow. Figure 2.1 provides a summary of

Feasibility
_____ PROJECT START

Engage commissioning authority.

Discuss commissioning scope/budget.

Develop preliminary Commissioning Plan.

Set up predesign commissioning team.

Develop (or review) Owner's Project Requirements.

Confirm commissioning scope/budget.

Finalize predesign Commissioning Plan.

Establish Issues Log format/process.

Ensure A/E incorporates Commissioning Plan in fee.
_____ PHASE TRANSITION

PREDESIGN PHASE

Set up design phase commissioning team.

Review/update OPR and CX Plan.

Develop expectations/scope/format for design phase documents.

Develop Basis of Design.

Develop Training Plan.

Develop Systems Manual requirements.

Develop Construction Checklists ("drafts").

Verify BOD and prepare to share with contractor.

Verify Training Plan.

Verify Systems Manual format/requirements.

Verify first-cut Construction Checklists (and test procedures).

Incorporate checklist, testing, and training requirements into specs.

Verify Construction Documents.

Resolve Issues Log items.

Update OPR, BOD, and Commissioning Plan.

Verify updated documents.
_____ PHASE TRANSITION

DESIGN PHASE

Figure 2.1   Flow chart outlining major activities in the commissioning process.

_____ PHASE TRANSITION

Set up construction phase commissioning team.

Establish construction phase procedures.

Finalize Construction Checklists.

Finalize test procedures.

Finalize Training Plan.

Finalize construction phase Commissioning Plan.

Verify submittals.

Work the Issues Log.

Use checklists and conduct tests.

Verify checklist and testing completion and results.

Conduct training.

Verify training.

Develop Systems Manual.

Verify Systems Manual.

Resolve Issues Log items.

Update OPR, BOD, checklists, test procedures, Systems Manual and CX Plan.

Verify updated documents.

CONSTRUCTION PHASE

_____ PHASE TRANSITION

Set up occupancy and operations commissioning team.

Establish occupancy and operations phase procedures.

Complete deferred Construction Checklists.

Complete deferred test procedures.

Conduct scheduled/deferred training.

Resolve Issues Log items.

Assist with punch list management.

Conduct warranty review.

Complete Systems Manual.

Conduct lessons-learned workshop.

Update and verify all commissioning process documents.

Develop final Commissioning Process Report.

OCCUPANCY AND OPERATIONS PHASE

_____ PHASE TRANSITION

Ongoing Commissioning?

Figure 2.1   (*Continued*)

the flow of major commissioning activities across the several project acquisition phases. The information in Figure 2.1 should be viewed as an outline—not as an all-encompassing task listing. In addition, although activities are listed in general chronological order, many tasks are conducted in parallel, or in time-disconnected steps, or in cyclic patterns. In other words, do not be too smitten by the simplification of a complex process that is presented in this flow chart.

INTERVIEW

Views on the Commissioning Process

Jeremiah (Jere) Lahey, AIA, Project Director, Department of Management Services, State of Florida

Q. What commissioning steps/procedures/reports/documents has your Department dealt with?

A. *We are involved in all aspects of commissioning from an owner's perspective—contract negotiations, scope and fee, verifying that benchmarks are met before payments are released, reviewing documents produced during the commissioning process, ensuring that comments and deficiencies noted by the commissioning authority are corrected, that field applications are noted in the commissioning reports, that Systems Manuals and training are provided to the maintenance staff.*

Q. Is there generally a cost or a cost savings due to use of the commissioning process?

A. *We see mixed results—cost savings are dependent on the commissioning authority, final owner of the building, engineer of record, and the contractor. The more sophisticated the building requirements (lab versus office), and the more knowledgeable the team members involved, the more there is an opportunity to find a sophisticated, innovative design solution that provides an energy-efficient building. The greatest cost savings resulted when commissioning was started during the design phase, and continued until training was completed and the building was handed over to the owner. The design/construction costs were increased by having a commissioning authority or team involved, but the cost savings were realized in the energy savings over time and the ability to more easily maintain the mechanical systems.*

Q. Did commissioning aid in the successful completion of the project? Or did commissioning cause project delays or encumber the design/construction/turnover of the project to the owner?

A. *This depended primarily on the commissioning authority, although the other team members could influence the outcome as well. The commissioning authority was responsible for the tone or atmosphere that the* comments and deficiencies were provided under. The owner needed to set the standard for how much risk insurance was required for the scope of design. Project delays could result from too much securitization. Savings in both time and money resulted when deficiencies were discovered early on and corrected.

Q. How was the commissioning process received by other members of the project team? Was it supported, or was there resistance?

A. The more sophisticated and knowledgeable the team members— namely the owner, engineer of record, contractor, and sometimes the architect—the more commissioning was embraced and viewed as a positive influence on the final outcome—an energy-efficient, easily maintained building. In other cases, commissioning was viewed as an intrusion by the engineering design team, causing delays resulting from interference with the design process. Some contractors appreciated the commissioning authority's review of completed tasks as building construction progressed, where the commissioning authority discovered deviations from the original scope that could have cost the contractor untold hours and expense if discovered at a later date and then corrected. Other contractors found this review process onerous.

Q. In your opinion, what types of projects lend themselves to commissioning?

A. Buildings that require complicated mechanical systems, or have specific and somewhat unique scope requirements (e.g., a crime lab) have a better opportunity to meet the needs of the owner if commissioning is started in the design phase.

Q. How do you think commissioning influences LEED projects?

A. LEED and commissioning go hand in hand. LEED recognizes the importance of mechanical systems having optimum energy performance, and commissioning provides the opportunity for greater success.

REFERENCES

ASHRAE. 2005. *ASHRAE Guideline 0: The Commissioning Process*. The American Society of Heating, Refrigerating and Air-Conditioning Engineers, Atlanta, GA.

Mills, E. 2004. "The Cost-Effectiveness of Commercial-Buildings Commissioning: A Meta-Analysis of Energy and Non-Energy Impacts in Existing Buildings and New Construction," Lawrence Berkeley National Laboratory, Berkeley, CA. http://eetd.lbl.gov/emills/PUBS/Cx-Costs-Benefits.html

Chapter 3

The Commissioning Team

TEAMWORK IS NECESSARY

Just as it is important to understand that commissioning is an ongoing process and not a short-term event, it is equally important to realize that commissioning is accomplished by a team and not by a lone-wolf expert or troubleshooter. Except for the simplest of buildings, it is unlikely that any single person would have the expertise required to successfully validate a reasonable range of design documents, equipment, assemblies, and systems. Even for a small building, it is highly unlikely that any one person could adequately represent the diverse views of the many stakeholders involved with the building acquisition process. Teamwork is necessary to provide the breadth and depth of interest and experience that will lead to a successful commissioning outcome.

The commissioning team will be led by an entity (an individual or firm) called the *commissioning authority*. In the past, the term commissioning *agent* was often used, but the word *agent* implies a legal ability to generally act on behalf of someone (in this case, the owner)—which is not the intent of the commissioning process. The commissioning authority is typically engaged by the owner under a professional services contract and will strive to further the owner's interests, but will normally not be empowered to speak on behalf of the owner. The exact nature of the owner-commissioning authority relationship will be spelled out in the professional services agreement between these two entities. This is, as might be imagined, an important document.

The commissioning authority will lead the commissioning team. The commissioning team will, at different times during the extended building-acquisition process, include various representatives of the owner, the design team, the contractor, subcontractors, and other specialists (as necessary or appropriate), along with representatives of the commissioning authority. The size and composition of the team will change over time—being most compact during predesign and then again during occupancy and operations and expanding greatly during construction. Some members of the commissioning team are specifically engaged to participate on the team as a primary responsibility (the commissioning authority), other members are assigned to the team by the owner, and yet other members (from the design team, contractors, specialists) must be brought onto the team through appropriate requirements included in broader contractual relationships.

A good commissioning team will be composed of people who want (or at least are paid to want) to participate, who are knowledgeable about their areas of responsibility, and who can make day-to-day decisions on behalf of the constituency they represent. The importance of the owner in developing a functional commissioning team cannot be overstated. The expected contributions of the owner, commissioning authority, design professionals, and contractor to the commissioning team are generally outlined immediately below. Specific members of the team are then further discussed.

THE COMMISSIONING AUTHORITY

The commissioning authority is the core of the commissioning team. Care in selecting this leadership entity is critical. As commissioning is a long-term process involving numerous parties with potentially conflicting interests, it can be argued that management skills are as important as technical skills when selecting a commissioning authority. Specific technical expertise can be brought onto the commissioning team as required, but the ability to communicate, negotiate, and deal fairly with people are characteristics that must reside with the commissioning entity.

There has been much debate over the issue of who can best provide commissioning services. Today, the majority of commissioning providers are engineering based, most commonly with a mechanical or electrical engineering background. This does not preclude architects or even owners from providing commissioning services. These two options, however, are rarely seen today. Repeated attempts to bring architects more actively into the commissioning arena in the mid-1990s were generally unsuccessful—most likely because commissioning was broadly (and mistakenly) viewed as fooling around with equipment rather than managing project quality. Owners with qualified building management staffs could also commission projects, although independence of opinion and action might be a legitimate concern for the design team and contractor.

Key Building Commissioning Organizations

The following organizations provide certification in commissioning and/or substantial resources related to the commissioning process.

ASHRAE: American Society of Heating, Refrigerating and Air-Conditioning Engineers, Inc. Atlanta, Georgia. www.ashrae.org

Certification in commissioning is under development (in 2008); publishes Guideline 0 and Guideline 1 dealing, respectively, with the commissioning process and HVAC&R technical commissioning requirements; developing Guidelines 1.2 and 1.3 to address commissioning of existing buildings/systems and training during commissioning.

BCA: Building Commissioning Association. Portland, Oregon. www.bcxa.org/

Membership is an indicator of commitment to commissioning and the commissioning process; provides certification of commissioning providers.

PECI: Portland Energy Conservation, Inc. Portland, Oregon. www.peci.org/

Long-term advocate of building commissioning. Convener of the annual national commissioning conference (a great information resource and networking opportunity); provider of substantial information on building commissioning.

CACX: California Commissioning Collaborative. Sacramento. www.cacx.org/

Source for a substantial and growing collection of information about commissioning.

ACG: Associated Air Balance Commissioning Group. Washington, DC. www.aabchq.com/commissioning/

Provides a certification exam to test basic knowledge and technical expertise of eligible commissioning providers. Publishes the *ACG Commissioning Guideline*.

NEBB: National Environmental Balancing Bureau, Gaithersburg, Maryland. www.nebb.org/

Provides for certification of firms and for the qualification of individuals through the Building Systems Commissioning Program (addressing HVAC, plumbing, and fire protection systems).

UW: University of Wisconsin-Madison, Department of Engineering Professional Development, Madison. http://epdweb.engr.wisc.edu/

Long-term provider of short courses on various aspects of commissioning; offers a certificate/certification program for commissioning providers.

Independence of action is a valuable asset for a commissioning authority. The commissioning authority will often act as a communications bridge between the various parties to the building acquisition process. Having this bridge be perceived as neutral can be very important to the success of such communications. Even more critically, the commissioning authority will occasionally have to make a determination that a design solution or equipment/assembly installation will not meet the Owner's Project Requirements or comply with the Contract Documents. It is absolutely critical that such a determination be seen as coming from a neutral party. Hiring an independent commissioning authority with no connections to the principal project parties is strongly recommended.

An independent commissioning authority must also be qualified to undertake a proposed project. Qualifications are best addressed via a robust Request for Qualifications/Request for Proposals process. Experience on similar projects (with successful outcomes) is likely to prove the best qualifier. A number of commissioning provider certification programs are currently in place or under development. Although the value of such certifications has not been clearly established, seeking a commissioning authority with diverse qualifications—including certification—is sensible. The commissioning authority should have experience with projects of similar scale and scope, and should be a member of one or more professional associations that support and promote commissioning. The commissioning authority should also be able to demonstrate familiarity with current commissioning process guidelines, should be appropriately licensed or certified, and should be truly interested in the project at hand. Seeking the lowest possible bid for commissioning services is not recommended as a terribly sensible approach.

Commissioning Guidelines

The following commissioning guidelines are generally available to assist various member of the commissioning team in developing and executing their roles in the commissioning process.

ASHRAE Guideline 0-2005: The Commissioning Process (American Society of Heating, Refrigerating and Air-Conditioning Engineers).
Generally considered to provide an appropriate framework under which to undertake commissioning of building systems and assemblies.

ASHRAE Guideline 1.1-2007: HVAC&R Technical Requirements for The Commissioning Process (American Society of Heating, Refrigerating and Air-Conditioning Engineers).
Provides updated guidance for the commissioning of HVAC&R systems in conformance with Guideline 0-2005.

ASHRAE Guideline 5-1994: Commissioning Smoke Management Systems (American Society of Heating, Refrigerating and Air-Conditioning Engineers).
Provides technical guidance for commissioning of these systems.

ASHRAE Guideline 1.2-20xx: The Commissioning Process for Existing HVAC&R Systems (American Society of Heating, Refrigerating and Air-Conditioning Engineers).
Under development; addresses retrocommissioning of HVAC&R systems.

ASHRAE Guideline 1.3-20xx: Building Operation and Maintenance Training for the HVAC&R Commissioning Process (American Society of Heating, Refrigerating and Air-Conditioning Engineers).
Under development; addresses the training aspects of commissioning with a focus on HVAC&R systems.

NEBB: *Procedural Standards for Building Systems Commissioning* (National Environmental Balancing Bureau).
Provides guidance for the development and implementation of a plan for commissioning of building HVAC and plumbing systems.

NEBB: *Design Phase Commissioning Handbook* (National Environmental Balancing Bureau).
A document intended to assist the commissioning professional in reviewing building mechanical system design documents.

SMACNA: *HVAC Systems—Commissioning Manual* (1994, Sheet Metal and Air Conditioning Contractors' National Association).

A guide for contractors, owners, and engineers interested in commissioning for new buildings and the recommissioning of existing buildings.

AABC: *ACG Commissioning Guideline* (2nd edition, Associated Air Balance Council).

This guide includes methodologies for various types of commissioning, the scope of services performed in commissioning, and sample forms.

NIBS Guideline 3-2006: Exterior Enclosure Technical Requirements for the Commissioning Process.

National Institute of Building Sciences. Provides technical guidance for the commissioning of building enclosure elements.

CACX: *California Commissioning Guide: New Buildings*. California Commissioning Collaborative.

Provides an overview of the commissioning process for new construction.

CACX: *California Commissioning Guide: Existing Buildings*. California Commissioning Collaborative.

Provides an overview of the commissioning process for existing buildings.

OWNER REPRESENTATIVES

The project owner or client will play an important role in the commissioning team throughout the life of a project—usually through the input of various representatives who will participate during the several project phases. Especially critical input from the owner/client will come during the development of the Owner's Project Requirements (OPR) document. This document will often be developed by the predesign commissioning team (although other approaches may be used), with active representation from the owner. Anything that can be done to improve the quality of the OPR by ensuring broad participation in its preparation is desirable. *ASHRAE Guideline 0* provides information on workshop techniques that have proven successful in eliciting broad-based input from stakeholders within an owner's organization during OPR development.

It is critical that an owner's operations and maintenance (O&M) staff participate on the commissioning team during preparation of the Owner's Project Requirements (to instill a sense of reasonableness about in-house capabilities to run and maintain systems) and during construction and occupancy and operations (as the project is turned over to the owner and responsibility assumed by the O&M staff). If in-house operations expertise is not available prior to project turnover, it is a good idea to bring in such expertise (familiar with the type of project and local conditions) via short-term contract. During design, operations representatives will be able to provide valuable input to the commissioning team relative to training needs and formats (which will be incorporated into the Contract Documents). During construction, commissioning team members representing the owner should be involved with performance verification activities and will be the audience for training efforts. During occupancy and operations, owners' representatives on the commissioning team will help to ensure a smooth facility turnover and informed planning for long-term facility use.

DESIGN TEAM REPRESENTATIVES

Ideally, representatives from the design team (architects, engineers, specialists) would participate on the commissioning team throughout all phases of a project. In practice, this will usually be difficult to arrange. It may be difficult or impossible to involve the design team during predesign phase meetings to assist with development of the Owner's Project Requirements and the first-cut Commissioning Plan because the design team has not yet been selected by the owner. It may be difficult to obtain design team participation during construction due to a current tendency by many AE firms to minimize construction phase project involvement. This can be addressed by appropriate language in the professional services agreement requiring designer participation on the commissioning team during construction. It will likely be even more difficult to obtain design team participation in occupancy and operations phase commissioning team meetings. This difficulty can also be overcome through appropriate contract language—and perhaps a reminder that participation at this stage of a project would be educational for the designers (a chance to pick up lessons to be learned for the team's next projects).

Active and appropriate participation on the commissioning team by the designers during the design phase of a project is mandatory for successful commissioning. Providing for verification of equipment, assembly, and systems installation and operation is a key element of building commissioning that must be included in the Contract Documents, so these activities can be bid, scheduled, and implemented by the contractor. The design team is responsible for preparation of these documents. Collaboration between the commissioning authority and representatives of the owner, the design team, and the contractor/subcontractors to develop effective and acceptable verification checklists and procedures will do a lot to turn commissioning tests into a cooperative effort to ensure quality rather than an outsider's attempt to find problems.

The design team will also be responsible for incorporating training requirements into the Contract Documents. Participation of the design team in the definition of such requirements will help to ensure that training is timely, meaningful, and can have long-term impact.

CONTRACTOR REPRESENTATIVES

The importance of having active contractor representation on the commissioning team cannot be overstated. Without such participation, commissioning process requirements will typically be seen as being imposed from outside (rather than as self-implemented), and the valuable construction experience residing in the contractor will not be usefully tapped to improve the commissioning process.

It would be great to have contractor representation during predesign-phase commissioning team meetings. This may be feasible if an integrated design approach to project acquisition is taken. It may be more difficult to do under traditional design-bid-build approaches. Contractor input at this stage of a project would allow the commissioning team to tap into valuable insight regarding costs and project scheduling during development of the Owner's Project Requirements and refinement of the Commissioning Plan.

A similar situation exists during the design phase. Contractor participation in the development of verification checklists (that will become part of the Contract Documents) should contribute to more workable checklists and test procedures. This is especially true since in most cases the contractor will be required to conduct the test procedures on behalf of the commissioning team. In lieu of contractor participation, an experienced commissioning authority should be able to bring past experiences into the development of viable test procedures for commonly encountered equipment, assemblies, and systems. Such experience is less likely to be available for unusual equipment and assemblies and/or unique systems or intersystem arrangements. In such cases, tapping into experiences from manufacturers or outside consultants may be valuable. The contractor (if available during the design phase) would also be expected to contribute to the development of a Training Plan for the owner's operations and maintenance staff.

Contractor participation on the commissioning team during the construction phase of a project is mandatory and must be ensured through appropriate language in the agreement between the owner and the contractor. Explicit language is recommended over an implicit assumption of participation. Contractor completion of construction tasks will be spelled out in the specifications that comprise part of the Construction Documents. Coordination activities for an unfamiliar process (commissioning), however, may not be properly bid without some guidance regarding expectations. The same concern about contractor participation is true of the occupancy and operations phase, when it may be difficult to obtain serious contractor buy-in due to project close out (except for warranty issues).

SPECIALISTS

Numerous specialists could beneficially contribute to the commissioning process through timely participation on the commissioning team. Such specialists, and their importance, will vary from project to project. Methods of ensuring their participation will also vary. In general, a clear Commissioning Plan and appropriate contract language can bring expertise into the commissioning team from specialists who would normally not participate. A key point to remember when it comes to specialists is that they are usually responsible for equipment, an assembly, or a system that is critical to project success, which is why they are involved in the first place. Bringing them into the commissioning process will be equally important for project success.

COMMISSIONING TEAM PARTICIPATION EXPECTATIONS

Tables 3.1 to 3.4 summarize the contributions expected from the most commonly involved members of the commissioning team during each of the project phases. The expectations are neither inclusive nor exclusive—smaller projects will see less involvement, larger projects greater involvement.

Table 3.1 Predesign Phase Commissioning Team Members and Roles

Representing	Participant	Expectations	Comments
Commissioning Authority	Commissioning authority (an individual or several people)	Act as lead author for development of initial Commissioning Plan; assist in development of Owner's Project Requirements (OPR).	Should be responsible for the Commissioning Plan (with input from owner); may be responsible for developing the Owner's Project Requirements document (depending on professional services agreement).
Owner	Owner	Define expectations and budget available for commissioning process; engage commissioning authority; assign appropriate staff to participate on commissioning team.	Owner initiates the commissioning process.
	Functional area representatives	Able to contribute information regarding the required/desired quality of facility to OPR discussions.	Critical to a good OPR.
	Operations and maintenance representatives	Able to contribute information about owner capabilities and expectations for facility operations and maintenance to OPR discussions.	Critical to a good OPR.

	Project manager	General coordination of owner contributions to the commissioning team.	Availability will depend on project scale and owner's capabilities and on project structure.
Design Team	Architect and mechanical engineer (at a minimum)	Provide insights into OPR feasibility; contribute to discussions regarding the Commissioning Plan.	Design team would benefit from participation by getting a heads-up on commissioning plans.
Contractor	Project manager	Provide insights into OPR cost and schedule ramifications; contribute to Commissioning Plan.	Availability will depend on project delivery approach that is chosen.
Specialists	As appropriate	Contribute to OPR; contribute to Commissioning Plan.	Need will be determined by specific project characteristics.

Table 3.2 Design Phase Commissioning Team Members and Roles

Representing	Participant	Expectations	Comments
Commissioning Authority	Commissioning authority (an individual or several people)	Update and expand the Commissioning Plan; review the Basis of Design; coordinate development of Construction Checklists; manage the Issues Log.	Generally responsible for maintaining currency of commissioning process documentation and for managing the flow of information among commissioning team members.
Owner	Owner	Review and approve updated Commissioning Plan; respond as necessary to issues raised in Issues Log; act as arbiter in the event of concerns about the impact of design decisions on OPR.	The Owner has ultimate approval and decision- making authority for all commissioning process documents; he/she may defer to the Commissioning Authority for technical issues (such as checklists).
	Functional area representatives	Review Basis of Design to assure that OPR is being properly addressed.	A last chance before construction to assure that needs have been addressed.
	Operations and maintenance representatives	Review Basis of Design to assure that institutional O&M needs and capabilities are being properly addressed during design; provide input to draft Training Plan.	Getting a heads-up on planning for equipment and systems testing may be very useful to Owner; staff needs must be considered in developing Training Plan.

	Project manager	General input to Commissioning Plan updates with respect to Owner's schedule and budget.	Contributions from this entity will vary with project roles and responsibilities.
Design Team	Architect and all key consultants	Develop Basis of Design; develop construction drawings that reflect the OPR; develop specifications that address the OPR and include all commissioning process activities (especially verification and training); develop or provide input to Construction Checklists; establish requirements for Systems Manual.	The Construction Documents and Basis of Design are the purview of the design team; development responsibility for Construction Checklists and Training Plan will vary, depending on team member roles as defined in their services contracts.

(continued)

Table 3.2 (*Continued*)

Representing	*Participant*	*Expectations*	*Comments*
Contractor	Project manager	Provide input to all commissioning process activities regarding cost and scheduling implications; such input is especially critical to development of viable Construction Checklists; coordinate information input from subcontractors and suppliers.	Although a contractor may not be on board during the design phase, a contractor's experience can be valuable in ensuring that commissioning requirements are doable and practical.
	Subcontractors	Provide input to development of Construction Checklists and Training Plan in area of expertise.	Can provide useful input regarding practicality and costing of proposed Construction Checklist requirements.
	Major equipment suppliers	Provide input to development of Construction Checklists and Training Plan in area of expertise.	Should be the most knowledgeable source regarding start-up, shake-down, and training requirements.
Specialists	As appropriate	Role is similar to that of suppliers.	Input varies with specialist's role.

Table 3.3 Construction Phase Commissioning Team Members and Roles

Representing	Participant	Expectations	Comments
Commissioning Authority	Commissioning authority	Act as point person for all commissioning documents; orchestrate contractor verifications (via checklists and tests); verify appropriate development of training and Systems Manual; maintain Issues Log; chair commissioning team meetings.	Commissioning authority ensures that the substantial commissioning effort undertaken in this phase goes smoothly and that information flows and is acted upon.
Owner	Owner	Will make decisions required to maintain integrity of the project and the commissioning process.	May be called on to revise OPR if in the best interests of the project.
	Operations and maintenance representatives	Will observe appropriate equipment and systems verifications and actively participate in training activities.	Will be gearing up for impending takeover and operation of the building.
	Project manager	General coordination of owner contributions to the commissioning process and team.	Availability will depend on project scale, owner's capabilities, and project structure.

(continued)

Table 3.3 *(Continued)*

Representing	Participant	Expectations	Comments
Design Team	Architect and consulting engineers	Will respond as required to field changes or substitutions that may affect OPR; provide training on Basis of Design as required.	Ongoing participation of design team members in the commissioning process should be secured through professional services contract.
Contractor	Project manager and others	Play a major role in executing commissioning activities—specifically completing Construction Checklists and systems tests; training owner's personnel, preparing the Systems Manual, and responding to Issues Log items.	Contractor participation must be defined in the Contract Documents (essentially the specifications). Verification and testing will typically be done by the contractor under the eyes of the commissioning authority and selected owner's staff.
	Subcontractors	Will conduct many of the verification and testing activities outlined above, as well as elements of training.	Requirements for involvement must be defined in the specifications.
	Major equipment suppliers	May be responsible for overseeing testing of key equipment and providing factory or site training.	Involvement will not occur without clear specifications.
Specialists	As appropriate	Role may be similar to that played by major equipment suppliers.	Involvement will need to be spelled out in professional services contract.

Table 3.4 Occupancy and Operations Phase Commissioning Team Members and Roles

Representing	Participant	Expectations	Comments
Commissioning Authority	Commissioning authority	Will coordinate close-out of the commissioning process—including completion of deferred verifications, tests, and training; resolution of Issues Log elements; and verification and delivery of all commissioning process documentation. May embark on an ongoing commissioning process for facility.	Although the commissioning process is often winding down during this phase, diligent completion of all activities is critical to a successful project (allowing the owner to assume knowledgeable use of the facility).
Owner	Owner	Will accept activities as completed to permit transition from an acquisition role to that of operator.	Close coordination with the commissioning authority is likely.
	Operations and maintenance representatives	Assume responsibility for facility operations; complete training; accept and begin using Systems Manual.	This hand-off will to a great extent establish the likelihood for successful facility operations.

(continued)

Table 3.4 (*Continued*)

Representing	Participant	Expectations	Comments
	Project manager	General coordination of owner contributions to the commissioning team and this transitional phase.	Availability will depend on project scale, owner's capabilities, and project structure.
Design Team	Participants as appropriate	Assist in bringing closure to the commissioning process as per the structure defined for the project—this may include training activities, completion of the Systems Manual, etc.	Typically, the design team is trying to move on at this point in the process; they need to be retained to provide such assistance in transitioning that is only available through their collective expertise.
Contractor	Project manager	Work to ensure that the contractor completes his/her extensive contributions to the commissioning process, with particular emphasis on deferred verification, testing, and training, Systems Manual completion, resolution of Issues Log concerns, and project turnover.	At this point, time and money are usually running out and there is a desire to move on. The commissioning team must ensure that the transition is smooth and viable.

	Subcontractors	Resolve Issues Log concerns and participate in deferred validation, testing, and training.
	Major equipment suppliers	As for subcontractors above
Specialists	As appropriate	Assist with activities such as training and performance benchmarking as defined in services agreements.	Because specialists are often involved with complex and mission-critical systems, their participation may be invaluable in ensuring a smooth transition to owner use of a facility.

INTERVIEW

Views on the Commissioning Process

Thomas E. Cappellin, P.E., Senior Mechanical Engineer, Hanson Professional Services Inc., West Palm Beach, Florida

Q. As a provider of commissioning services, what key benefits of commissioning do you see repeatedly?

A. *Key benefits include:*

- *When design documents are reviewed by the Commissioning Team during the design phase many coordination issues are discovered and addressed, which aids in reducing design conflicts and omissions.*
- *When design phase issues of conflict and omission are resolved the project will usually experience a reduction in contractor requests for information (RFIs) submitted during the construction phase. Most RFI activity results from inaccurate, inconsistent, and missing design-phase documentation.*
- *Owner's operating personnel are introduced to their new building systems by receiving a structured series of formal training sessions that are organized and monitored by the commissioning authority. In addition they receive Record Documents and Systems Manuals that are complete and specific to the equipment, assemblies, and systems that they will operate and maintain during the building's occupancy.*

Q. In your opinion, is commissioning worth its cost—in terms of benefits to the owner? Why?

A. *Applying the commissioning process to a building, new or renovated, is definitely a benefit to the owner in terms of reducing, or eliminating, many unplanned and unforeseen costs that are encountered during the construction and occupancy phases. In most cases, these additional costs can exceed the cost of the commissioning process.*

Q. What seems to be the biggest misperception regarding commissioning among owners?

A. *Owners may expect commissioning services to address all elements, equipment, and assemblies that are a part of, or installed in, the building. This perception can cause a breakdown in the owner's relationship with the commissioning team. It is imperative that the scope of commissioning services be clearly documented and understood by all parties prior to acceptance of an agreement for commissioning services.*

Q. What seems to be the biggest misperception regarding commissioning among contractors?

A. Contractors not familiar with the commissioning process may consider it to be an extra layer of supervision over their work that will cause them additional cost and delay in their contractual obligations. They may be resentful and suspicious of the commissioning team and be uncooperative in working with it. Over a period of time, this resentment can be reduced or removed if the commissioning team is proactive in convincing the contractor that it can help to reduce his or her costs by discovering, and resolving, conflicts with building elements or other trades early in the construction phase.

Q. What do you recommend as a means of breaking the ice on commissioning—for the novice owner?

A. A novice owner needs to be provided with a concise, but complete, explanation of why the commissioning process was developed and how it can be of benefit to his or her building project. This can be augmented by providing case histories of previous projects and reasons why the commissioning process was a success, or why it failed expectations.

Q. Is there anything that the commissioning professional needs to get better at—to improve outcomes for an owner?

A. The commissioning professional needs to be sensitive to owner expectations and maintain a consistent line of communication. The commissioning professional must also provide the owner's team, design team, and construction team with reports and documentation on a timely basis to enable the efficient flow of information and successful on-time completion of the project.

Q. There is a linkage between green building design and commissioning. Is this a good connection? Do you see areas of potential improvement?

A. Green building design is becoming the path of choice for many government and sophisticated building owners. Therefore, those involved in commissioning activities need to understand green building design concepts and apply that knowledge during reviews made during the design phase and observations made during the construction phase. The most satisfying result of commissioning is knowing that this activity has enabled a building to meet or exceed its energy and sustainability goals established during the conception and design phases.

Q. Do you have any words of wisdom for those newly embarking on the building commissioning process? Or for seasoned professionals?

A. At the top of my "lessons learned" list is a lesson learned early in my commissioning activities. The owner needs to be provided with credible reporting and documentation during all phases of the project. The documentation should include accurate meeting minutes, complete descriptions of observations made, and resolution results for design phase *and construction phase issues. The commissioning authority provides written work to all parties involved in the project. Therefore, all deliverables must be clearly understood, accurate, and timely, so that progress of the project is not delayed. The results of the commissioning process must show a positive effect on the project and provide the owner with a high comfort level and justify his or her choice of including commissioning in the project budget.*

Chapter 4

Commissioning Coordination

THE ROLE OF COORDINATION

Coordination is one of the principal features of the commissioning process. In fact, commissioning could be described as a series of verification activities preceded by extensive coordination efforts and archived by defined documentation requirements. Details of the documentation and verification activities underlying commissioning are presented in other chapters. This chapter focuses on coordination and associated communications.

A key fear of commissioning, expressed by many contractors (and, to a somewhat lesser extent, by design professionals), is that it will bog down the construction (or design) process by adding unnecessary paperwork, inspections, and potential choke points—all to the detriment of profits and schedule. The need for verifications was discussed earlier; looking at a few unverified (uncommissioned) projects, as in the cited Florida Solar Energy Center study, is the best possible argument for commissioning. The concern regarding paperwork and schedules bears consideration, but should not act to dissuade an owner from commissioning. The key to minimizing negative impacts on costs and schedules is the implementation of predictable, consistent, predefined coordination opportunities facilitated by smoothly functioning communications channels.

Coordination is sorely needed in the fragmented world of building design and construction, where truly effective coordination is often more a myth than a reality. The commissioning process can greatly assist in improving project coordination through effective communication and documentation. Coordination among all parties to the building acquisition process is important for successful project outcomes and is critical for high-performance buildings (those using passive systems, seeking green certification, pushing the envelope on energy efficiency, attempting to reduce carbon emissions, and/or with demanding functional requirements). The commissioning process provides a structure for such coordination.

There are literally hundreds of individuals involved in the design, construction, and operation of a typical building. These individuals tend to aggregate into groups affiliated with one or another of the key project players—owner, designer, contractor, operator. Communications (and its potential corollary—coordination) are typically strongest within a given group; although this does not necessarily mean it is always effective. Communication and coordination are typically weakest between groups. Formal communications channels, as are developed through the commissioning process, can improve coordination across the board, but especially between groups. This chapter is structured around several important coordination "events."

DEFINING AND CONVEYING PROJECT REQUIREMENTS

The starting point for the nuts and bolts of the commissioning process, and for establishing coordination channels, is the Owner's Project Requirements (OPR) document. Starting a project without a very strong OPR is like going to the grocery store with an incomplete shopping list—there will be regrets once back home (or upon moving into the new facility). The first task for the commissioning team (as distinct from the commissioning authority) is assisting in the development of the OPR document. This effort, properly done, leads to the following coordination links:

- *Between the owner and the commissioning authority.* This will be done on a regular working (versus simply a contractual) basis.
- *Between the owner and representatives of the owner's various user group constituencies.* These constituencies could include, for example, marketing professionals, in-office support staff, IT staff, behind-the-scenes support staff, and management.
- *Between the owner and the owner's facility operating staff.* This might include contract staff.
- *Between the commissioning authority and the owner's several constituencies.*
- *Ideally, between the owner and a design team representative.* This linkage may need to come later, depending on the project acquisition process in place.
- *Ideally, between the owner, design team representative, and a contractor's representative.* This might also need to wait.

The establishment and operation of the commissioning team at this formative stage in the project acquisition process can represent the first moves in setting up an integrated building design process. Integrated building design involves more than just a functioning predesign commissioning team, but the commissioning process can be a useful initiation into a better way of doing things (Elvin 2007). The coordination and collaboration links established in developing the Owner's Project Requirements should truly lead to all parties being on the same page—working toward a common (and well-thought-out) set of objectives. This is the essence of coordination: defined by Wiktionary as "making different people or things work together for a goal or effect." That different people are involved is evident; the different "things" are often the different cultures and ways of thinking held by each of the individuals or groups, which can seriously impede clear communications.

Annex I of *ASHRAE Guideline 0* provides recommendations for workshop facilitation techniques that can ease the development of complete and well-thought-out project requirements through involvement of a wide range of constituencies. Annex J of the guideline outlines key elements typically found in an Owner's Project Requirements document, including a very detailed matrix showing concerns to be addressed and a suggested organizational structure (ASHRAE 2005).

The preparation and early dissemination of the formal Owner's Project Requirements document furthers coordination via the following links:

- *Between the owner and the design team.* Not that they won't communicate, but structured, well-reasoned (and down-on-paper) communications can be a real asset in trying to describe something as complex as expectations for a building project (no matter the project scale).
- *Between the commissioning authority and the design team.* This connection will develop further as the project progresses.

DEFINING AND VERIFYING DESIGN SOLUTIONS

One of three key coordination drivers during the design phase is the establishment and conveyance of project design solutions in the form of Construction Documents. The commissioning process establishes formal communication channels and coordination activities to ensure that all necessary parties are fully in the loop during this phase of building acquisition. Although not formally a part of the Construction Documents, the Owner's Project Requirements and the Basis of Design assume important roles in the search for project clarity and coordination. Specific coordination linkages result from the development of design solutions:

- *Between the owner (and the owner's constituencies) and the design team.* The link comes through the intermediating power of the Owner's Project Requirements (OPR) document and its use to verify design decisions.
- *Between the design team and the commissioning team as the Basis of Design is developed and verified.* The owner's constituencies are an important player in this coordination.
- *Between the commissioning authority and the design team during verification of design documents against the OPR.*
- *Between various members of the commissioning team as the projected outcomes from design decisions are compared to previously defined project expectations.* Remember that this is a diverse group including representatives from all important parties.
- *Between the design team and the owner's operations personnel.* The implications of design decisions for operations and maintenance will be compared to the defined capabilities of the group that will assume responsibility for project success upon occupancy.
- *Ideally, between the design team and the contractor.* Availability of the Basis of Design provides an insight into the thinking underlying proposed project solutions.

DEFINING AND CONVEYING CONSTRUCTION VERIFICATION REQUIREMENTS

The second key coordination-causing activity during the design phase is the development and dissemination of criteria and procedures for the equipment, systems, and assembly verifications that will occur during the construction phase via the application of Construction Checklists and system/assembly testing procedures. The end result of good coordination on this effort will be a complete and unambiguous set of specifications that addresses verification activities to be conducted by the contractor (and the roles that others will play in these activities). Coordination linkages that will be developed during this effort include the following:

- *Between the design team and the contractor and subcontractors.* This is the ideal situation, but again depends on the project delivery model—otherwise the verification requirements being established will not benefit from contractor experience and expertise.
- *Between the design team and the commissioning authority.*

- *Between the commissioning authority and the owner's operating personnel.* This is relative to participation in verification activities.
- *Between the commissioning authority and (ideally) the contractor and key subcontractors.*
- *Between the design team, commissioning authority, and (ideally) specialists who may be involved with verification efforts.* This might include equipment manufacturers and test and balance contactors.

DEFINING AND CONVEYING TRAINING REQUIREMENTS

The third key coordination-causing activity during the design phase is the development and dissemination of training requirements for the owner's personnel (building operators and users). This effort will result in the development of a viable and clearly enunciated Training Plan and will involve these linkages:

- *Between the owner and the owner's operating personnel and the commissioning authority.* These linkages help define specific training needs and targets.
- *Between the commissioning authority and the design team.* This allows those involved to assess and define "design" information to be included in training.
- *Ideally, between the owner, commissioning authority, and contractor.* The purpose of this linkage is to assess and define "construction" information to be included in training.
- Ideally, between the owner, commissioning authority, contractor, and equipment suppliers. This allows those involved to assess and define "product" information to be included in training.

DEFINING AND CONVEYING OPERATIONAL INFORMATION

During the construction phase, coordination will occur during regular meetings of the commissioning team and during the numerous collaborative efforts to verify the installation and operation of building equipment, systems, and assemblies. In addition, planning for the handover of the project from the contractor to the owner sets the stage for the following coordination linkages:

- *Between the commissioning authority, design team, contractor, and owner.* This occurs as the Systems Manual is assembled and verified.

- *Between the design team, contractor, suppliers, and commissioning authority.* This occurs as training of owner's personnel is conducted.
- *Between the commissioning authority, owner, and contractor as punch list issues are dealt with.* In theory, this effort is minimized by effective use of the Issues Log.
- *Between the commissioning authority and the owner.* This occurs as the transition from construction to the occupancy and operations phase is planned and then implemented.

PLANNING FOR ONGOING COMMISSIONING

As the typically structured new building commissioning process reaches its end near contractor warranty expiration, discussions—that should always occur—about continuing the commissioning process under the auspices of an ongoing commissioning program provide an opportunity for further coordination:

- *Between the owner and his/her operations personnel.* This linkage is regarding long-term facility expectations and needs.
- *Between the commissioning authority and operating personnel with respect to future efforts to maintain project performance.* This includes energy efficiency, environmental quality, maintainability, flexibility, productivity, and overall value.

The beauty of the commissioning process may well lie in its ability to get a group of knowledgeable people (with a generally common big-picture objective, but often very divergent short-term goals) together at scheduled intervals to collaborate on completing defined tasks. Good things will happen under such a scenario. The commissioning process, as outlined by *ASHRAE Guideline 0*, sets up just such a situation, wherein coordination must and will occur. Where communications are effective, good things will happen. Where opportunities for both structured and random sharing of information are provided on a regular basis, good things will happen.

The ever-evolving Commissioning Plan provides a roadmap of overall expectations for coordination on a project. Other commissioning process activities, however, will have more of an impact on coordination in the trenches. Regularly scheduled commissioning team meetings, which start in predesign and continue through occupancy and operations, are the primary vehicle for structured (yet potentially informal and informative) communications and coordination. It is not possible to overemphasize the potential for a skilled commissioning authority, when chairing these meetings, to literally breed coordination and mutual respect among all players.

The Issues Log provides a more formal (but necessary) forum for communication and coordination. A well-run Issues Log serves as a written and regularly reviewed "to-do" list for all parties to the commissioning process. Thus, commissioning team meetings and the Issues Log are crucial to successful day-to-day coordination. Commissioning process reports serve a secondary coordination function. Not to denigrate the need for these reports, but they represent after-the-fact records of coordination more than real-time opportunities for coordination.

An interesting side question of "how much" communication with the owner is appropriate draws a range of responses from commissioning providers. Some feel strongly that the most successful commissioning implementation is one that runs totally in the background as far as the owner is concerned. This approach is probably not the best way to promote the benefits of commissioning (and its expenses) and/or obtain future projects from the same owner. It is good to not involve the owner in every decision, but surely, the owner should get a sense of action and results commensurate to the fee being paid for the commissioning service. To hope that a positive final commissioning report will tell the whole story of success is wishful thinking. It is likely much better to tell the story of small but accumulating successes visibly, but unobtrusively, as they occur during the course of a project.

Figure 4.1 summarizes key coordination focal points that will occur during the commissioning process and suggests the parties most likely to be involved with (and benefit from) the resulting coordination. This summary illustration is not intended to be all-inclusive of coordination requirements or roles.

INTERVIEW

Views on the Commissioning Process

Tim Corbett, Project Manager, Social Security Administration, Baltimore, Maryland

Q. As a consumer of commissioning services, what key benefits of commissioning do you see repeatedly?

A. Better organization and coordination of the work. The commissioning process gets the information on the owner's project requirements down to the level where the work is actually being done.

Q. In your opinion, is commissioning worth its cost—in terms of benefits to the owner? Why?

A. At 2 percent to 4 percent of total project cost for commissioning, it is a bargain. Quality Assurance/Quality Control (QA/QC) procedures for complex work may increase project overhead by 10 percent. To increase the value of commissioning, it needs to be coordinated and incorporated into the QA/QC requirements in Division 1 of the project specifications. The commissioning and project QA/QC efforts should complement each other instead of working as separate processes.

Q. What seems to be the biggest misperception regarding commissioning among owners?

A. Understanding the applications and differences between what CSI (the Construction Specifications Institute) calls the two types of commissioning. These being, whole building commissioning beginning at project inception, and what CSI calls traditional commissioning. Traditional commissioning usually takes place near the end of the construction phase and refers only to systems and components.

Q. What seems to be the biggest misperception regarding commissioning among design professionals?

A. They fail to see the value or the need for systems and equipment suppliers and installers to understand what is being built and why.

Q. What seems to be the biggest misperception regarding commissioning among contractors?

A. Looking at commissioning as another series of inspections and witch hunting.

Q. What do you recommend as a means of breaking the ice on commissioning—for the novice owner?

A. Keep the process to the most basic and understandable application. Emphasize how the process is aimed at preventing mistakes, not finding them after the fact.

Q. Is there anything that the commissioning profession needs to get better at—to improve outcomes for an owner?

A. It needs to be proactive in keeping the commissioning process transparent to the workflow and apparent to the vendors and installing contractors.

Q. Do you have any words of wisdom for those newly embarking on the building commissioning process? Or for seasoned professionals?

A. Maintain a positive working relationship with the work force, vendors, and owner.

DEFINE OWNER'S PROJECT REQUIREMENTS

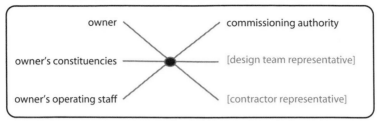

[commissioning team]

DISSEMINATE OWNER'S PROJECT REQUIREMENTS

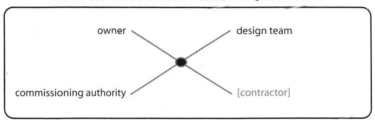

DEFINE/VERIFY DESIGN SOLUTIONS

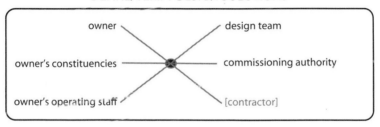

[commissioning team]

DEFINE/CONVEY VERIFICATION REQUIREMENTS

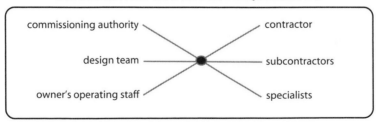

[commissioning team]

Figure 4.1   Coordination-rich efforts that occur during the commissioning process.

DEFINE/CONVEY TRAINING REQUIREMENTS

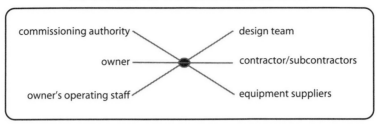

[commissioning team]

DEFINE/CONVEY OPERATIONAL INFORMATION

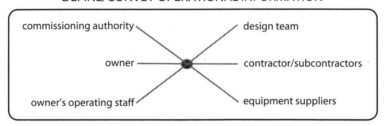

[commissioning team]

PLAN FOR ONGOING COMMISSIONING

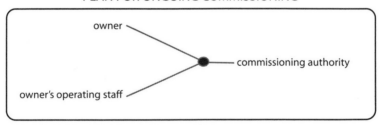

Figure 4.1   (**Continued**)

REFERENCES

ASHRAE. 2005. *ASHRAE Guideline 0: The Commissioning Process*. The American Society of Heating, Refrigerating and Air-Conditioning Engineers, Atlanta, GA.

Elvin, G. 2007. *Integrated Practice in Architecture*. John Wiley & Sons, Hoboken, NJ.

Wiktionary. www.wiktionary.org/ (accessed 2008).

Chapter 5

Verification and Testing

THE IMPORTANCE OF VERIFICATION

Communications, documentation, and verification are the defining characteristics of the building commissioning process. It would be easy to make the argument that verification is the heart and soul of commissioning and that communications and documentation simply support this central element. That argument will not be actively disputed. On the one hand, timely, appropriate, and focused verification is an irrevocable aspect of the commissioning process. On the other hand, running verification checks without the project-long, projectwide context provided by the other elements of the commissioning process is not recommended. Testing building components in temporal and spatial isolation is essentially contractor quality control—which should be done by the contractor as part of the construction process.

Before getting into the specifics of verification and testing, some discussion of terms is necessary. *ASHRAE Guideline 0* defines *verification* as follows:

> The process by which specific documents, components, equipment, assemblies, systems, and interfaces among systems are confirmed to comply with the criteria described in the Owner's Project Requirements. (ASHRAE 2005)

Guideline 0 defines *acceptance* as follows:

> A formal action, taken by a person with appropriate authority ... to declare that some aspect of the project meets defined requirements, thus permitting subsequent activities to proceed. (ASHRAE 2005)

In general, the commissioning process presumes that the commissioning authority (acting through the commissioning team) verifies, while the owner (perhaps acting through a project manager) accepts. Ideally, the commissioning authority will act as a professional and independent advisor to the owner regarding the appropriateness of technical work. An owner may choose to accept work (design or construction) that does not pass verification checks or tests. That is the owner's inherent right as the client/customer. The commissioning authority, however, will ideally explain the implications of such acceptance on the owner's objectives for a project.

Note that the ASHRAE definition of verification sets the Owner's Project Requirements (OPR) as the reference benchmark. This will raise no issues during design phase verification—when the essential question is whether or not the design approach and documents can deliver the outcomes required by the owner as expressed in the OPR. During construction, the situation is not so straightforward, as it is possible that some element of a project will meet the contractual requirements embodied in the Construction Documents, but not be able to deliver the owner's expected outcomes. This may occur if a disconnect between design and OPR has occurred—or when a change or substitution was permitted without a full understanding of all its implications. Such a situation will need to be evaluated by the owner and a decision made to change the work to meet the OPR (very likely at additional cost) or to accept the work and compromise the OPR (possibly also at additional, but deferred, costs related to reduced productivity or increased energy use).

The following discussion presents verification activities in a chronological manner focused on project phases. It would also be reasonable to look at verification through a focus on documentation. Although project phases have been selected as the means of organization, documents are interwoven into the discussion. The terms *verification* and *testing* will be encountered later in this chapter. These are not-quite-synonymous terms, in that testing generally implies a choreographed series of verifications, often focused on an environmental condition (such as thermal comfort, daylighting, response to an emergency condition, and the like). Testing typically involves looking at a range of conditions produced under a variety of scenarios. Verifications are often (but not always) yes-no decisions.

PREDESIGN PHASE

During the predesign phase, several foundation-setting documents are developed by various parties and accepted by the owner. These documents set the stage for the commissioning process and begin to address verification activities.

The Owner's Project Requirements are the established benchmark against which all verifications are made. Thus, this document needs to be as complete, thorough, and specific as to intended outcomes as possible. Critical performance aspects not addressed in the OPR will need to be evaluated against some measure—most likely a default value demanded by code, standard, or general practice, which may not necessarily track with the owner's (unfortunately unstated) desires. Quantifiable and measurable criteria should be established whenever feasible. This does not preclude qualitative criteria ("joints will appear uniform"), but if an outcome can be expressed in measurable terms ("minimum initial illuminance of 500 lux") verification and testing may be simplified. One pattern for OPR development is to first state design intent (providing a general direction for an outcome— such as "high energy efficiency") and to then benchmark the intent with a defined criterion (e.g., "30% better energy efficiency than ASHRAE Standard 90.1"). An owner or user can usually express intent quite well, but may need assistance from the commissioning team to match the intent with appropriate numerical (or other) design targets.

The Owner's Project Requirements documentation should be verified by the commissioning team (under the direction of the commissioning authority) and accepted (if in fact acceptable) by the owner. Is the OPR complete; is it practical; does the content match the owner's plain-language statements of intent? Verification of this particular document will draw on the commissioning authority's experiences— and all other project verifications will draw on the OPR.

Verification is a concern coloring many issues during the predesign phase. The initial Commissioning Plan will outline the scope, schedule, and budget for commissioning activities. Reasonable provisions for verifications, especially those occurring during the construction phase, should be worked into this first-cut plan. A format and procedures for effective use of the Issues Log process should be established. The Issues Log provides a formal mechanism for the recording, archiving, and resolving of deviations or concerns identified by members of the commissioning team. Many of the issues so identified will likely be verification-related.

Verification of project artifacts other than the Owner's Project Requirements will begin in earnest during the design phase. The requirements, expectations, and procedures for such verification activities— participation in which is not now the norm among the building design professions—must be communicated to those who will be involved. For example, the A/E team must be told in advance that its work will be subject to verification against the OPR, that design team members will be required to participate in and support such efforts (clearly within some boundaries of reasonableness—the intent is not to second-guess all design decisions), and participants must be provided with some sense of process (e.g., that random sampling will be used to select elements to be verified). Procedures for resolution of conflicts related to design verification should be made known. Typically, this may often be an implementation of the owner declaring "the bucks start and stop here."

The commissioning process transition from the predesign to the design phase will be marked by the owner's formal acceptance of the Commissioning Plan and Owner's Project Requirements. Such acceptance will be predicated on a recommendation to do so from the commissioning authority (following verification). The commissioning authority should also follow up with the owner to assure that appropriate information regarding commissioning process verification activities is conveyed to the design team (ideally, as part of a written agreement for professional services).

INTERVIEW

Views on the Commissioning Process

Jeff J. Traylor, CxAP, CIT, LEED-AP, Senior Engineer, EMCOR Government Services, Durham, North Carolina

Q. As a provider of commissioning services, what key benefits of commissioning do you see repeatedly?

A. *Shorter punch lists, better coordination between crafts, better training, and higher energy efficiency.*

Q. Is there any downside to the commissioning process as commonly practiced?

A. *None, if the process follows ASHRAE Guideline 0-2005. However, incomplete commissioning (not starting until construction is complete or nearly so) frequently gives results that do not meet expectations, thereby giving the whole process a bad reputation.*

Q. Has the development and publication of *ASHRAE Guideline 0* visibly improved commissioning services?

A. *Yes! It has created a standard of practice that clearly defines the process.*

Q. What seems to be the biggest misperception regarding commissioning among owners?

A. *That it starts at the end of construction.*

Q. What seems to be the biggest misperception regarding commissioning among contractors?

A. *That it interferes with their work processes.*

Q. What do you recommend as a means of breaking the ice on commissioning—for the novice owner?

A. *Start with a small project and give the commissioning authority some latitude.*

Q. Is there anything that the commissioning profession needs to get better at—to improve outcomes for an owner?

A. *Following the process and getting an OPR (Owner's Project Requirements) as the first step. Without an OPR, the commissioning process has no benchmarks to follow.*

Q. What do you see on the horizon for building commissioning over the next 5 to 10 years?

A. *If the industry acts responsibly, and does not fall prey to trying to do the work for too small a fee, it will thrive. Otherwise, it will sink into oblivion.*

Q. Do you have any words of wisdom for those newly embarking on the building commissioning process? Or for seasoned professionals?

A. *Learn the process—follow the process. Shortcuts will lead to failures and failures will have a long-lasting impact. Remember that one "Oh Shoot" will wipe out 100 "Att'a'boys"!*

DESIGN PHASE

During the design phase, the design approach and documents developed by the design team will be verified against the Owner's Project Requirements by the commissioning team. The primary intent of this verification is to assure that the design proposal can reasonably deliver the outcomes expected by the owner. To reemphasize, the objective is not to second-guess the design team. In no way should commissioning verification recommendations replace or supersede the legal and ethical responsibility for making and defending design decisions held by the designers of record. A secondary intent of design verification is to ensure that the Construction Documents are of a quality and at a level of completion that can support successful delivery of the owner's project expectations.

To the extent possible, design reviews should be accomplished incrementally, so that a concern that may permeate the entire approach or document set can be identified and corrected early to minimize schedule and cost impacts. A simple example of such a situation would be inconsistent or incomplete equipment identification/labeling that would make the use of Construction Checklists and the Issues Log much more difficult. Four distinct aspects of design verification are described in Guideline 0:

1. A broad-perspective review of Construction Documents for general quality. This includes completion, appropriateness, readability, cross-references, etc.

2. A broad-perspective review of the documents for interdisciplinary coordination.
3. A discipline-focused review (e.g., HVAC&R systems) for conformance with the Owner's Project Requirements.
4. A review of the project specifications for clarity, applicability, and conformance with the OPR and Basis of Design. Particular attention should be paid to the adequacy and appropriateness of the commissioning process aspects of the specifications (including training requirements).

As with the extensive verification efforts that will occur during the construction phase, it is recommended that design phase verification be done using a random sampling approach. In essence, this means selecting (admittedly somewhat arbitrarily, but without letting bias slip into the decision) a small percentage of the documentation for review and evaluation. A range of 10 to 20 percent of an element (drawing page, drawing set, specifications pages) is suggested as appropriate. Annex N of *ASHRAE Guideline 0* provides examples of sampling approaches for the commissioning process (ASHRAE 2005).

It is worth stating again that it is not intended that design verification substitute for quality control procedures implemented by the design team. The purpose of this review is to identify systematic issues that will negatively impact the likelihood of successful project outcomes. Sampling should accomplish this goal in that it is statistically unlikely that systematic issues uncovered during review will occur only in the sample area selected for review—or will occur elsewhere but not in the sample (and thus remain undiscovered).

Resolution procedures for identified design concerns will vary from project to project and with the group dynamics of the various parties and personalities involved. In an ideal world, all members of the commissioning team will be seeking the highest possible quality commensurate with profitability, and design verifications will go smoothly with appropriate give and take. In a less ideal world, the owner may be asked to act as arbiter between irreconcilable parties—in which case the focus of concern is achievement of the owner's documented requirements.

The commissioning team should also verify the Basis of Design document developed by the design team. Although this will not become a legally binding document, it will be of immense benefit to the commissioning team during the construction phase and to the owner's personnel during occupancy and operations. The OPR provides the benchmark for Basis of Design verification.

A major undertaking during the design phase will be the preparation of verification forms and test procedures to be used during the upcoming construction and ensuing occupancy and operations phases. Construction Checklists and test procedures/forms that are initially developed (drafted, outlined) during the design phase will play a key role in these efforts. Training requirements will also be identified and specified during the design phase for implementation during construction. Verification of the effectiveness of such training should be addressed. The responsibility for costs associated with retesting and/or retraining that is required due to verification failures must be contractually defined and made clear to all involved parties. A generally adopted pattern is that retesting costs are the responsibility of the party causing the "failure."

CONSTRUCTION PHASE

A substantial proportion of the commissioning process verification effort will occur during the construction phase. Construction Checklists and test procedures/forms play an integral role in streamlining and rationalizing these diverse activities. Drafted during the design phase, checklists and tests will typically be finalized during construction as the specific equipment, systems, and assemblies bid by the contractor become known. The operating philosophy of checklists and tests is that they should be specific to what is actually installed, be unambiguous, and require as little on-site interpretation/annotation/modification as possible. Time invested in planning for implementation will be repaid in reduced time for implementation.

The term *functional performance test* was (and still is) commonly heard in discussions about commissioning. *ASHRAE Guideline 0* does not use this phrase. This is partly due to baggage collected by this term over years of use and partly due to the much broader view of verifications presented in the guideline. Verification for the commissioning process is much more than confirming "function."

Although discussed under Construction Checklists in the chapter on documentation, an overview of the role of Construction Checklists in facilitating verification during construction is warranted. For each element (an item of equipment or component of an assembly) being verified, checklists will typically address the following:

- Safe delivery of the correct equipment/component: What was specified and approved was actually delivered in undamaged condition.
- Preinstallation condition: After sitting on site for some period of time and being moved around, the equipment/component is still undamaged.

- Quality of installation: The correct element was correctly installed in the correct location. At this stage of verification, reference to the Owner's Project Requirements becomes very important—a "normal" standard of installation care (say ductwork tightness) may not be acceptable on a project seeking high energy efficiency.
- Proper operation of the element in isolation: Does the equipment or component operate as a stand-alone element as intended and expected.
- Negative issues encountered: Ideally the construction checklists will be completed with the majority of "checks" falling in the "yes" boxes; when this is not the case a clear description of the deviation must be noted, along with an anticipated corrective action (this event should become part of the Issues Log).
- Linkages to test data forms related to benchmarking equipment performance and/or its operation as part of a larger system.

A narrative example of the use of Construction Checklists for an air diffuser will illustrate this process. Upon delivery, the diffusers are confirmed to match those specified and approved via submittals. The diffusers are noted as being in good condition. When distributed to the various rooms for installation, several diffusers are seen to have been damaged in storage (one has bent louver blades, several have scratched finishes); this is noted on the checklists and corrective action requested (give and take among the members of the commissioning team may decide whether replacement, repair, or installation as-is is most appropriate). A sampling of diffusers following installation shows that all that were looked at were properly installed in the correct locations (ductwork connections are tight and diffuser fit into the ceiling grid is good). A TAB (test and balance) report shows that the sampled diffusers perform as specified (adequate throw, acceptable noise generation, and expected airflow delivery). For this particular component, there may be a linkage to a systems test intended to verify that the HVAC system collectively can provide the thermal conditions called for in *ASHRAE Standard 55* (adopted as a performance benchmark by the Owner's Project Requirements).

It is anticipated that the commissioning team will be involved with verification of submittals (again using a statistical sampling approach), and with observations of tests or inspections required by code or other regulatory/supervisory bodies. Seasonal testing (verifying system performance to OPR criteria) may be accomplished during construction or may need to be deferred until appropriate weather conditions are available. Verification of the Systems Manual, as developed to date, should be a part of the transition from construction to occupancy. The same is true of verification of training conducted during the construction phase. There should be few, if any, serious unresolved issues sitting in the Issues Log at the end of construction. Verifying that this is the case should help to greatly reduce (if not eliminate) owner's punch list items. The collaborative work of the commissioning team during the construction phase should greatly ease transition to occupancy.

The bottom line for construction phase verifications is to collectively verify by sampling (as opposed to guaranteeing through 100 percent testing) that the project is ready for hand-off to the occupancy and operations phase. The objective is to reasonably ensure that the building or facility is ready for the owner to effectively occupy and/or use.

Most of the verification occurring during the construction phase will take place behind the scenes and not involve the owner. The owner, however, will be asked to accept the updated Basis of Design, the updated Commissioning Plan, and any revisions/updates to the Owner's Project Requirements as a closure to the construction phase. Unresolved Issues Log concerns may be referred to the owner for action.

OCCUPANCY AND OPERATIONS PHASE

This phase of project acquisition begins at substantial completion and is recommended to continue through the end of the contracted warranty/correction period for the facility and its major dynamic components. The majority of activity will usually be concentrated near the front end of this period. The broad purpose of the commissioning process during this phase is to verify that the facility as delivered and operated will meet the Owner's Project Requirements (as updated and revised). Continuation of commissioning throughout the life of the facility (ongoing commissioning—see Chapter 8) is highly recommended.

In summary, verification efforts during this phase will seek to ensure the following:

- Deferred (either intentionally or unintentionally) tests are successfully conducted.
- Training scheduled for this phase is successfully completed.
- The Systems Manual is both complete and acceptable.
- The Owner's Project Requirements and Basis of Design reflect any modifications necessitated by events occurring during this phase.
- Equipment performance and systems operations do not substantially degrade between initial occupancy and the end of warranties.
- The owner can assume long-term operation of the facility with adequate and appropriate documentation that will ease long-term decision making.

At the end of this phase the owner will be asked to accept the Final Commissioning Process Report—which should detail verification of all testing and training conducted during the phase, the successful completion of the Systems Manual, and the successful resolution of the Issues Log. Lessons learned from the process should be included in the final report.

Table 5.1 summarizes key verification activities by project phase and indicates important documentation linkages.

INTERVIEW

Views on the Commissioning Process

Kristin Heinemeier, Western Cooling Efficiency Center, University of California Davis, Davis, California

Q. As a long-term advocate for building commissioning, how would you describe the landscape for commissioning services today (as opposed to, say, 10 years ago)?

A. *I believe that there is more interest in energy efficiency now than there ever has been, and also more interest in generally operating buildings more cost-effectively. These trends help encourage interest in commissioning. There are more consultants involved in the building process, so it seems less of an extravagance to hire a commissioning expert. Of course, LEED is increasing the prevalence of commissioning, though I'm not sure if it's raising general awareness much.*

Q. What is the most positive thing about today's commissioning environment?

A. *Anything that tends to make people question how their buildings are working or look more closely at their building delivery process is a good thing.*

Q. Are there any negatives about today's commissioning environment?

A. *As the market for commissioning expands, with demand growing more quickly than the number of suppliers, I fear that the quality of commissioning may be suffering. Especially with LEED, where people aren't at all motivated to get the results, but just to get the commissioning in there.*

Q. What seems to be the biggest misperception regarding commissioning among owners?

A. *I think commissioning requires active involvement by the owner. It's not something they can just buy and forget about. To get the most out of it, they have to be involved.*

Q. What seems to be the biggest misperception regarding commissioning among design professionals?

A. I think design professionals see commissioning as the "design police," with somebody out there trying to find ways that the designer screwed up. They may also see commissioning professionals as invading their turf. They don't tend to see the commissioning professional as a part of their team.

Q. What seems to be the biggest misperception regarding commissioning among contractors?

A. I think contractors also see commissioning providers as cops looking for trouble. It's important for the commissioning provider to see it as their job to make the contractor look good, not look bad.

Q. What do you recommend as a means of breaking the ice on commissioning—for the novice owner?

A. Maybe get involved in retrocommissioning of an existing facility. Once they see the opportunities for improving the building, they'll be motivated to catch those at the start. Hire a commissioning provider to participate in a renovation, or smaller project before tackling a large project.

Q. What developments do you see on the horizon for building commissioning over the next 5 to 10 years?

A. I think that commissioning will bifurcate into a low-cost commodity service and a high-value advanced engineering service. The tools, objectives, processes, expected benefits, etc., will be very different for these two tracks, and the better we can understand and communicate the differences, the less confusion there will be in the market. Productivity tools for commissioning professionals will be an important development, to bring the cost down a bit.

Q. What key resources would you recommend to an owner or design firm about to embark on building commissioning?

A. Guideline 0, of course! The California Commissioning Collaborative has a lot of good resources (case studies, guides, tools—www.cacx.org/). I think their guide to commissioning of new buildings is a good start (and it's not just because I was a primary author!). Energy Design Resources (EDR) also has a lot of good resources. The commissioning assistant on the EDR Web site is a good tool—www.energydesignresources.com/.

Table 5.1 Outline of Key Verification Activities and Related Documentation

Project Phase	Key Verification Activities	Key Documentation Links
Predesign	Include a preliminary sense of verification elements in the commissioning schedule and budget.	The preliminary Commissioning Plan will address verification activities.
	Develop owner's project requirements with verifiable outcomes.	Owner's Project Requirements
	Establish format and procedures for Issues Log.	Issues Log format and structure
	Establish verification process and forms to be used for design verification.	Construction Checklist formats
	Verify the OPR prior to acceptance.	An accepted Owner's Project Requirements
Design	Develop verification forms and procedures for use during the construction phase: – Checklists – Test procedures	"Draft" Construction Checklists "Draft" test procedures and forms
	Verify design approach and design documents.	Construction Documents (general documentation; specific system documentation; and specifications)
	Establish verification process and forms for use with training.	Training Plan
	Implement Issues Log usage.	Operational Issues Log

74

Construction	Use Construction Checklists as a verification mechanism.	Final Construction Checklists
	Implement test procedures as part of the verification process.	Final test data forms
	Verify effectiveness of training.	Final training verification forms
	Verify appropriateness of Systems Manual.	Almost-complete Systems Manual
	Use Issues Log to track and resolve verification concerns.	An active Issues Log
Occupancy and Operations	Complete all Construction Checklists.	Completed and archived Construction Checklists
	Complete all test procedures.	Completed and archived test data forms
	Complete verification of training.	Completed and archived training verification forms
	Complete verification of Systems Manual.	Completed Systems Manual
	Close out Issues Log.	Generally resolved Issues Log

Note: The continually evolving Commissioning Plan will be linked to most of the above activities, but is not explicitly so noted.

REFERENCE

ASHRAE. 2005. *ASHRAE Guideline 0: The Commissioning Process*. The American Society of Heating, Refrigerating and Air-Conditioning Engineers, Atlanta, GA.

Chapter 6

Documentation

COMMISSIONING DOCUMENTATION

Commissioning—a process that attempts to ensure that quality outcomes are produced by the project design, construction, and operation sequence—is in essence about the communication and verification of expectations. The documentation described in this chapter supports both of these objectives. Some of the documentation described herein will be found in virtually every facility design-construct-operate situation. Such documents are independent of the commissioning process—although they are useful for commissioning. Other documentation discussed below is specific to commissioned projects. Although developing meaningful documentation consumes resources (time and money), the short- and long-term value of good documentation cannot be overstated. Without appropriate documentation things get missed, information is lost, and vital communications can become ineffective or even counterproductive. The purpose of this chapter is to give an overview of the documentation elements (and related communications) that are an inherent part of the commissioning process. These documents, when developed with care, will give the owner a better product and better value.

THE COMMISSIONING PLAN

The Commissioning Plan is the roadmap for the commissioning process. It grows and evolves as a project progresses and provides the owner with a record of the commissioning process at the end of a project. The Commissioning Plan is an absolutely essential document. At the start of a project, the Commissioning Plan will describe the intent for commissioning in general terms—clearly defining the expected scope of commissioning efforts and establishing a budget for the commissioning process. At this early stage, the Commissioning Plan answers the question: What does commissioning mean for this project? As a project moves through subsequent phases, the Commissioning Plan is expanded to address questions of who is involved and when, and what their responsibilities are.

At any point in the commissioning process the Commissioning Plan must provide a detailed vision of commissioning activities scheduled for the next project phase and a general description of activities that will occur in later project phases. With the incorporation of appendices (such as the Owner's Project Requirements, Basis of Design, Issues Log, Meeting Minutes, etc.) the Commissioning Plan also provides a detailed archival record of commissioning activities over the life of a project. Annex G of *ASHRAE Guideline 0* provides a suggested table of contents for a prototypical Commissioning Plan (ASHRAE 2005).

The commissioning authority, with input from the members of the commissioning team, is typically responsible for developing the Commissioning Plan. The commissioning authority and the owner will have primary input early in the process. The design team will become more involved during the design phase, and the contractor more involved during design (if possible) and during construction.

OWNER'S PROJECT REQUIREMENTS

The document known as the Owner's Project Requirements (OPR) is the foundation of the commissioning process. This formal and structured document describes what—from the owner's perspective—will constitute a successful project. Traditional building programs or "briefs" typically address spatial requirements (floor areas and adjacencies, for example) but rarely provide explicit requirements for the expected quality of spaces, especially relative to environmental conditions (thermal, visual, acoustical comfort; and their controllability). The OPR also provides an opportunity for the owner to document expectations and assumptions regarding issues such as system reliability, in-house maintenance and operations capabilities, and energy and environmental performance.

The purpose of the OPR is to establish a comprehensive benchmark for expected outcomes that will serve as the reference for design and verification efforts throughout the design, construction, and occupancy sequence. During design, the fundamental commissioning process question is: Will the proposed assembly, equipment, or system be able to provide the owner's stated performance objectives? During construction, the question shifts slightly: Does this installed assembly, equipment, or system deliver the owner's required performance? During occupancy, the question shifts a bit more: Does this assembly, equipment, or system as currently used or operated meet the owner's expressed performance needs? The OPR, updated throughout the life of a project to reflect current needs and realities, provides the reference against which these questions are asked.

Responsibility for development of the Owner's Project Requirements will vary from project to project in response to various contractual agreements for services. Whatever the arrangement put in place, lead responsibility must be made clear and the input of required parties provided for in a timely manner. Assisting with development of the OPR is the first key role for the commissioning team, and provides a great opportunity for the team to come together as a functioning and supportive entity.

BASIS OF DESIGN

The Basis of Design document is prepared in order to provide a written and logically structured record of the thinking that lies behind the design team's solutions as presented to the contractor. The OPR document represents a problem to be solved by the design team. The Construction Documents represent the team's chosen solution to that problem. A complex and substantial decision-making effort goes into developing that solution—most of which, as a historical record, is often poorly documented and seldom made available to parties outside of the design office. The Basis of Design allows this important information to be made available to the commissioning team and to be passed on to the owner for use in running the completed facility. This places design conditions, assumptions, equipment selections, operational narratives, and control sequences in the hands of the commissioning team (who can use this information to validate achievement of the OPR) and the owner's operating and maintenance personnel (who can use it to better operate the facility for the owner).

ASHRAE Guideline 0 (in both the main body and in Annex K) provides an extensive list of information that should be presented in a Basis of Design (ASHRAE, 2005). In summary, this includes the following: a list of governing codes and applicable standards and guidelines; key design criteria (performance targets); calculation methods and tools used; assumptions used in system selection and sizing; narrative explanations of system and equipment choices; a listing of equipment manufacturers and models used to establish aspects of the design; narrative explanations of control strategics; any limits or directives imposed by the owner; and assumptions regarding owner operating and maintenance capabilities.

The Basis of Design (BoD) should not be a part of the Construction Documents—but should be made available to the contractor as supplemental information. The Basis of Design will become a part of the Systems Manual. The potential uses for this collection of information are numerous. The BoD may assist the contractor in evaluating the appropriateness of a proposed change of equipment; it may assist the owner in understanding what additional plug loads can be supported by the HVAC system's cooling capacity; it may assist the operating staff in understanding the extent of spare pumping capacity or what is supposed to happen to the smoke control system during a power failure. The Basis of Design can be very useful in tracking system performance over time and reacting to drifts in performance parameters.

Responsibility for development of the Basis of Design must lie with the design team. The designers are the ones making the decisions leading to the basis of design, and they are the only ones who can capture and organize the information informing such decisions. As this is not a typical design product for most designers, the need to produce a Basis of Design must be called out in the professional services agreement between the owner and the design team.

CONTRACT DOCUMENTS/CONSTRUCTION DOCUMENTS

The Contract Documents include all the legal agreements between an owner and a contractor necessary to get a building constructed. The Construction Documents (mainly the project drawings and specifications) are the heart of the Contract Documents. These are required for all projects and are not uniquely commissioning process documents. The key aspect of these documents in a commissioned project is that they must address the commissioning efforts that are required of the contractor such that they can be included in the contractor's bid for the project. Anything not so included will likely be seen as a change-order add-on (unless the contractor is a philanthropist). Most, if not all, commissioning activities that involve the contractor will be described in the specifications (versus the drawings). The specifications must do a good job of defining the contractor's commissioning responsibilities.

Project specifications will typically address commissioning requirements, both broadly (under the general conditions) and specifically (under the individual sections describing the various building systems). In both locations, the requirements must be described in such a way that a rational bid can be prepared. This may require the use of allowances for activities (such as verification tests) that cannot be fully detailed at the time the specifications are prepared. An experienced commissioning authority can be of assistance to the design team in preparing such allowances (and the commissioning requirements in general). Likewise, a design team with commissioning process experience will more easily be able to prepare specifications that appropriately convey realistic contractor requirements for project commissioning. There are "guide" specifications available from several sources to assist with the preparation of useful commissioning sections for a project.

Responsibility for development of specifications that will lead to informed (and properly budgeted) contractor involvement with commissioning activities lies with the design team. The commissioning authority and commissioning team can provide assistance with this task, but the design team has contractual and professional responsibility for this important documentation. Annex L of *ASHRAE Guideline 0* provides a narrative discussion of the rationale behind commissioning process specifications, as well as a sample structure for such specifications (ASHRAE 2005).

CONSTRUCTION CHECKLISTS

ASHRAE Guideline 0 promotes the use of Construction Checklists as a means of focusing the commissioning process during the latter part of the design phase and during the critical construction phase upon a logical sequence of equipment, system, and assembly verifications. Verification, as previously noted, is defined as follows: "The process by which specific documents, components, equipment, assemblies, systems, and interfaces among systems are confirmed to comply with the criteria described in the Owner's Project Requirements" (ASHRAE 2005). Previously used and commonly heard verification-related terms such as *prefunctional testing* and *functional testing* are not used in Guideline 0.

Checklists may be used in other phases of the commissioning process, but will find their full potential during the construction phase. Development of effective and efficient Construction Checklists is a key commissioning activity during the design phase. Ideally, checklists to be used on a project will be included in the Construction Documents as a means of conveying the contractor's expected scope of commissioning verification efforts. Where this is not possible due to uncertainty about equipment or assembly selections, a draft checklist can provide an outline of intended effort.

The basic purpose of a Construction Checklist is to provide a script for verification efforts, combined with an easily completed written trail of such efforts. Checklists should anticipate positive responses—requiring elaboration only to describe problem situations. They should be customized for a specific project and, to the extent possible, be completed by literally checking a box or inserting a measurement. The completed checklists will be valuable in focusing discussions regarding noncomplying situations during commissioning team meetings and will serve as a useful benchmarking record for ongoing facility operations when included in the Systems Manual.

The following discussion should help to explain the evolving role of a checklist during construction. A lighting fixture is used as an example. The word *may* is used to reinforce the idea that sampling for verification is recommended (and, thus, certain elements may not be selected for verification):

- The checklist may note that the fixture was looked at during design verification and was considered in compliance with the Owner's Project Requirements (perhaps relative to a limit on luminance related to a desire for low glare potential).
- The checklist may note that the submittals (shop drawings) conformed to the fixture specified in the Construction Documents.
- The checklist may verify that the correct fixtures were delivered to the job site, were in good condition when delivered, and were properly stored.
- The checklist may verify that the fixture was installed in the correct location and was still in good condition.
- The checklist may verify that the light fixture was equipped with the correct ballast and the correct lamps were installed.
- The checklist may verify that the fixture operated correctly when connected to the electrical system.
- The checklist may verify that the fixture operated correctly under local switching control.
- The checklist may verify that the fixture properly dimmed under control of a daylight compensation control system.
- The checklist may confirm that the fixture behaved as anticipated in an emergency power outage situation.

The essential idea suggested in this example is simple. It makes no sense to wait to verify fixture performance as part of a daylighting system, only to discover that the fixture was damaged in shipping and had the wrong ballast installed. Options for all parties are limited at the end of the construction process. An issue that could have been easily corrected earlier becomes a serious hassle later. Skillful use of checklists can support the commissioning process as it attempts to avoid such unnecessary situations.

There is no single correct way to assign responsibility for development of Construction Checklists outside of the context of a specific project. Who should take the lead in preparing these documents will vary with participant capabilities and with the various professional services contracts that are in place. In any event, these roles must be spelled out as early as possible so this responsibility does not become an unwanted (or unaccepted) surprise. It is strongly recommended that the commissioning team be used to refine draft checklists developed by appropriate parties. Trying to write a checklist by committee is not usually a great idea, but not soliciting interested-party input on an activity they will be involved with is also not a great idea. This is an area where a well-functioning commissioning team can be a particularly valuable asset.

Annex M of *ASHRAE Guideline 0* provides generic recommendations for the types and organization of Construction Checklists (ASHRAE 2005). ASHRAE Guideline 1.1-2007 provides extensive sample checklists—with a focus on typical HVAC&R system components. A sample testing procedure to verify compliance with the Owner's Project Requirements is also provided in Guideline 1.1 (ASHRAE 2007).

TRAINING PLAN

The Training Plan is a subset of the Commissioning Plan that lays out the expectations and requirements for the training of owner's operating and maintenance personnel that will accompany the commissioning process. The intent of this training is to provide the types of information that will allow the owner to make a smooth transition from purchaser to user. As with the parent Commissioning Plan, the Training Plan should clearly define the scope of required training, outline the anticipated training activities and schedule (including deferred training), and provide a clear understanding of who is responsible for each training activity. Annex P of *ASHRAE Guideline 0* provides a sense of what should be addressed in a Training Plan (ASHRAE 2005).

The collective insights of the commissioning team will again be a valuable resource as this plan is developed. Requirements for training presented in the Training Plan must be conveyed to the contractor (and subcontractors and suppliers) through the Construction Documents. As with other commissioning activities that may not be fully resolved at the time of project bidding, allowances can be used to ensure that known training elements will not be overlooked even though not fully defined. Some training will likely be the responsibility of the design team (e.g., Basis of Design review) or the commissioning authority (providing an overview of the Owner's Project Requirements). Training by these parties falls outside of the Construction Documents and must be addressed in the respective professional services contracts.

Responsibility for development of the Training Plan falls to the commissioning team (lead by the commissioning authority). The Training Plan will typically be a subsection of the Commissioning Plan developed by these same entities.

SYSTEMS MANUAL

The term *Systems Manual* was adopted by *ASHRAE Guideline 0*, in lieu of the more commonly used Operations and Maintenance (O&M) Manual nomenclature, as a means of emphasizing that this documentation is focused around the building systems. The Systems Manual is intended to include the materials commonly found in an O&M Manual—as well as additional materials developed during the commissioning process. All the included materials need to be relevant to the systems actually in place, be organized to afford easy accessibility, and be useful to someone trying to ensure that the Owner's Project Requirements can be met throughout the occupancy and operations phase of a project. The typical O&M Manual is often not up to this task; a good Systems Manual should be.

ASHRAE Guideline 0, within the body of the guideline and more specifically in Annex O, recommends that the following elements be included in the Systems Manual—and not just be thrown in, but integrated such that relevant information is easily found:

- The Owner's Project Requirements
- Basis of Design
- Record Drawings
- Submittal documents
- Construction Checklists
- Test procedures and collected data
- Training information and records
- Operations information
- Maintenance information

This information should ideally be linked to supporting process documentation such as the Issues Log and commissioning team Meeting Minutes. The fundamental idea is to provide those responsible for delivering the Owner's Project Requirements on a daily basis all the information necessary for them to make informed decisions about equipment, systems, and assemblies as they are used, wander in performance or response, require repair or adjustment—and to do so in a manner that reduces access times and enhances the potential for success. This describes a document that is skillfully assembled by a caring professional who understands the needs of the ultimate user.

Responsibility for development of the Systems Manual will vary from project to project in response to various contractual agreements for services. Just about every party to the commissioning process (owner, design team, contractor, suppliers, TAB specialists, etc.) will have some responsibility for input. The key to a successful document is to have a lead "author" for this important product. Whatever arrangement is put in place, lead responsibility must be made clear and the timely input of required parties must be provided for via contract.

ISSUES LOG

Effective communications and documentation are two of the hallmarks of a successful commissioning process. The Issues Log for a project knits these two concerns together in a tool that can expedite the identification and resolution of problems. In today's computer environment, use of an electronic format for the Issues Log is highly recommended. The basic idea of this documentation is to provide a formal, easily accessible repository for concerns raised by various members of the commissioning team during the commissioning process—and their eventual resolution (including impacts on other commissioning documents).

The specifics of the Issues Log format and procedures should be developed early in the commissioning process and should be communicated to all members of the commissioning team who will participate in its use. It is possible that some team members will have write access to the log, while other may only view the entries. It is critical that the authenticity of entries not be compromised by the access procedures. As with Construction Checklists, the simpler the log entry procedures (while maintaining usefulness), the better. An example of how an Issues Log might be used follows:

- During design verification, the commissioning authority notes a concern about the ability of a diffuser to meet noise criteria established in the Owner's Project Requirements and posts the concern to the Issues Log.
- The design team engages the issue (flagged to their attention), reviews the selection, determines that a diffuser that would meet the criteria would cost three times more than the proposed diffuser, and notes this on the Issues Log.

- The commissioning authority retrieves the design team's response, determines that a change in criteria would probably be in the owner's best interest, and suggests changing the OPR to reflect this option.
- The owner accepts the recommendation and a note is made to change the OPR.

Underlying the Issues Log is a desire to have a transparent, collaborative, and cooperative doorway to problem identification, discussion, and resolution. Issues that cannot be resolved without face-to-face interaction between team members would be placed on the agenda for a commissioning team meeting. Items should not fall through the cracks, as it is easy to "pull up" unresolved issues quickly—along with complete documentation of each issue. At the end of a project, the Issues Log can serve as a useful record to show what the commissioning process accomplished by way of problem avoidance/resolution.

Responsibility for development and maintenance of the Issues Log throughout the commissioning process should lie with the commissioning authority.

MEETING MINUTES

Although mundane compared to most of the previously discussed documents, the preparation and dissemination of accurate and timely commissioning team Meeting Minutes is an important communications tool. The minutes should be prepared, disseminated, and archived by the commissioning authority and should be included in interim and final Commissioning Process Reports.

COMMISSIONING PROCESS REPORTS

Final and interim Commissioning Process Reports are an integral part of the documentation of the commissioning process. They provide commissioning team members with a current picture of commissioning process status, as well as easy access to an organized record showing the evolution of commissioning activities. These reports should be succinct to facilitate communications, but should reference other commissioning documents as appropriate.

Responsibility for preparation, dissemination, and archiving of Commissioning Process Reports lies with the commissioning authority.

Figure 6.1 provides a schematic summary of the development stages of major commissioning process documents during the various phases of a

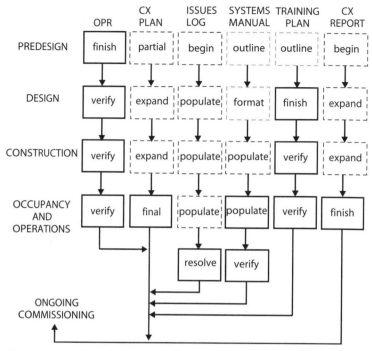

Figure 6.1 Development of key commissioning documents during the course of the commissioning process.

project. The intent of this figure is to give a sense of the give and take of the documentation efforts that accompany the commissioning process. Annex D of *ASHRAE Guideline 0* provides a more detailed (but tabular) description of document development schedule and responsibilities (ASHRAE 2005).

REFERENCES

ASHRAE. 2005. *ASHRAE Guideline 0: The Commissioning Process*. The American Society of Heating, Refrigerating and Air-Conditioning Engineers, Atlanta, GA.

ASHRAE. 2007. *ASHRAE Guideline 1.1-2007: HVAC&R Technical Requirements for the Commissioning Process*. American Society of Heating, Refrigerating and Air-Conditioning Engineers, Inc., Atlanta, GA.

Chapter 7

Training

TRAINING OWNER'S PERSONNEL

The training aspects of the commissioning process are focused on providing the owner's facility management personnel—particularly those involved with facility operations and maintenance—with practical knowledge of installed building systems and their characteristics. The objective of this training is to permit such personnel to effectively employ the systems to meet the Owner's Project Requirements during the life of the facility. Providing these critical people with such information is not the norm in current practice. The commissioning process is structured to improve this situation. The training aspect of commissioning will be a consideration in every project.

The training of building occupants/users on aspects of building operation that they might substantially influence is also a part of the training process. An example of the latter would be found in most buildings with passive systems or those striving for high performance. For example, occupant control of blinds can have a serious impact on daylighting performance. Occupant-controlled plug loads can account for a major portion of the electrical load in a building—and rational control of plug loads can affect the ability of a PV system to displace grid-supplied electricity.

Training may be provided by the contractor, by subcontractors, by manufacturer's representatives, by the design team, and/or other members of the commissioning team. Training may occur in the factory, on the building site (where equipment is installed), in the classroom, or via the Internet. Training may be conducted using live instructors or pre-recorded media, sit-down lectures, or hands-on experiences. The keys to successful training are that it be preplanned, well-executed, instructive, and effective. Training may address both generic and project-specific concerns as appropriate—providing, for example, an overview of green roof performance expectations as well as the intent, design, and maintenance requirements for a specific roof.

ASHRAE Guideline 0 provides general recommendations for the training elements of the commissioning process. Annex P in the guideline presents a sample training agenda (ASHRAE 2005). An ASHRAE guideline project committee (GPC 1.3) is currently working to develop a separate guideline to address training during commissioning in greater detail.

THE TRAINING PLAN

The Training Plan, an element of the broader Commissioning Plan, is the evolving roadmap for the training aspects of the commissioning process. It will grow and increase in specificity as a project progresses. When the owner assumes use of the project, the Training Plan will provide an archive of training materials that can be used throughout the life of a facility. The Training Plan is a critical document that should be given appropriate attention during all phases of the commissioning process. At the start of a project, the Training Plan will describe the expectations for the training aspects of commissioning—while clearly defining the expected scope of training efforts, establishing a budget for these efforts, and setting up a preliminary schedule for training activities. During the construction phase, the Training Plan must provide a clear and specific indication of who will be trained, on what systems, by whom, using what proposed methods. Training activities during the various phases of a project are discussed in this chapter.

PREDESIGN

During this initial project phase, development of the Owner's Project Requirements document should be used as a catalyst to establish needs and expectations for training of facility personnel and users. Workshops used to define the outcomes that will characterize a successful project can also be used to establish training requirements. Questions to answer during this process include: what are the capabilities of the owner's personnel, what general information about likely building systems would be important to know (problematic systems, new systems, new approaches to operations and maintenance), what specific information about likely systems would be of most use to personnel and users, what role should/might users play in ensuring successful systems operations? If the owner will be contracting for operations and maintenance services, rather than using in-house personnel, it would be advantageous to include a service representative on the pre-design commissioning team.

At the end of the predesign phase, a draft Training Plan should be developed and included in the Commissioning Plan. The Training Plan should indicate the anticipated scope of training activities and identify the roles various participants are expected to play to accomplish the necessary training effort. It is critical that training to be conducted by members of the design team (e.g., providing a review of the Basis of Design) be included in the professional services contract for design services. As with any unconventional service element, it is not wise to assume that the service provider will freely and ungrudgingly participate. Explicit (to the extent possible) is better than implicit (hoping it will happen) when it comes to such expectations. Some sense of how verification of training will be handled during the commissioning process might also be worth considering during this phase—for example, will checklists be used (as with other aspects of verification)?

DESIGN

The general objective of addressing training during the design phase is to refine expectations for training while ensuring that training activities that are to be provided by the contractor are included in the Construction Documents in a manner that permits rational bidding. This will help to avoid training being seen as an unbudgeted burden being imposed on the contractor—ideally leading to better and more effective training experiences. The commissioning team must establish the details of training:

- What systems, equipment, assemblies will be the focus of training
- What those to be trained know about such systems—what their current capabilities are
- What information needs to be conveyed to assist the personnel in meeting the Owner's Project Requirements
- How such information can best be conveyed in an effective manner
- The specifics of the type, provider, location, duration, and outcomes of training sessions
- Estimated times and schedules for training sessions

Defined outcomes should be established for each training session— probably easiest to do through written objectives and content outlines.

A means of reasonably verifying the effectiveness of training should be developed. As with other aspects of verification, use of a sampling approach is recommended.

Most training will typically be conducted during construction so that the owner's personnel are prepared to assume operation of a facility immediately upon turnover. This intent must be conveyed to the providers and receivers of the training so that mutually agreeable scheduling can be set up. Some training may logically spill over into the occupancy and operations phase, but this should be minimized and only involve activities best done during this later phase.

Training should address both the general and the specific. The general will provide a context for the specific. General information to be conveyed would typically include an overview of the facility, key aspects of the Owner's Project Requirements, and a summary of critical elements of the Basis of Design. The specifics to be addressed during training are described in *ASHRAE Guideline 0* (ASHRAE 2005):

- Instructions regarding operations during emergency situations (and addressing the various types of emergencies anticipated)
- Information to assist in day-to-day operations, with an eye to maintaining performance in accord with the Owner's Project Requirements
- Troubleshooting guidance
- Adjustment guidance for equipment and systems likely to wander from installed performance, with benchmarking data from verification testing
- Day-to-day maintenance procedures, presented in a manner that meshes with the owner's preferred maintenance strategy (failure, predictive, etc.)
- Repair procedures (including assumptions for replacement and removal)
- Updating of the Systems Manual and associated recordkeeping

The development of training documentation must be addressed during the design phase. The Systems Manual, to be prepared during the construction phase, will be both a source of training materials and will archive other training artifacts. The benefits of coordination of efforts should be considered when allocating responsibility for these various documentation efforts. Guideline 0 suggests that a Training Manual be prepared as a record of training activities (ASHRAE 2005). This manual would contain the Training Plan, training agendas and syllabi, training materials (handouts, videos, DVDs, etc.), and evaluation instruments and results. Links to specific contents in the Systems Manual that will assist in ongoing training for owner's personnel should be provided.

To ensure that training activities are actually and satisfactorily carried out, the contractor's training responsibilities need to be included in the project specifications (typically in Division 1). To the extent possible, specific requirements (factory-training, hands-on learning experiences, classroom sessions) should be spelled out. Where uncertainty exists regarding equipment selections and the like, allowances may be included. Any such allowances should be adequate to permit the quality of training demanded by the Owner's Project Requirements. Expectations for the development of training artifacts above and beyond the norm (such as building system- or operation-specific DVDs) should be explicitly noted.

CONSTRUCTION

The majority of training will take place during the construction phase in an attempt to bring the owner's personnel up to speed on systems, equipment, and assemblies before they assume responsibility for operations and maintenance upon project turnover. To assist in making this flurry of activity go smoothly, training requirements should be discussed in any prebid meetings (along with other aspects of the commissioning process). The commissioning team, and regular team meetings, can be used to facilitate training during construction. Means to maximize the effectiveness of training while reducing its time demands can be considered. It might be possible, for example, to overlap on-site systems training with owner personnel observation of performance verification testing.

The construction phase is a busy and often chaotic time for all involved with the building acquisition process. It is important that training not be lost in the shuffle. Training effectiveness should also be verified as soon as reasonable after elements have been completed so that ineffective training incidents can be rectified. As with other verifications, owner recourse to deal with undelivered or poorly delivered training episodes should be spelled out in the project specifications. Failure of training to meet the Owner's Project Requirements, while meeting the letter of the Construction Documents, is a possibility. Options for addressing such a situation should be in place. Proper documentation of commissioning training must be part of the construction phase activities—documentation (archiving, indexing, cross-referencing) should not be deferred to some ill-defined "later" time.

OCCUPANCY AND OPERATIONS

Some training will naturally and desirably be deferred until an owner has assumed possession of a facility and has begun day-to-day use. In addition, some training intended to occur during the construction phase may be deferred for unforeseen, but justifiable, reasons. Although the contractor may be anxious to move out and onward, it is important that these remaining elements of training be conducted and be conducted well. Training for facility occupants/users will typically take place during this phase—although some such training could occur during construction if the project involves a move of existing users from one facility to another.

The general intent of training for owner's personnel is to ease and facilitate the transition from buyer to user, while enabling long-term, most-beneficial-use of a facility. To accomplish these objectives it is reasonable that there be a training blitz near the point of project turnover. It is also to be expected that training will occur throughout the life of a facility. New employees will come on board and need to be familiarized with specific systems, equipment, and assemblies. Established employees may benefit from refresher training. New occupants may need to be introduced to unusual or performance-enhancing features of a building (lighting occupancy sensors, as an example). Provisions should be made to ensure that such ongoing training is not just possible, but is likely and convenient. It is not reasonable to tell a new employee to "read the Systems Manual" just to become acquainted with operational expectations for a heat recovery ventilation system. Training aimed at occupants should be supported with materials that make sense—a pamphlet or Web-based fact sheet, for example, not a technical manual.

Table 7.1 provides an overview of the training activities and documents that are considered an integral part of the building commissioning process.

REFERENCE

ASHRAE. 2005. *ASHRAE Guideline 0: The Commissioning Process.* The American Society of Heating, Refrigerating and Air-Conditioning Engineers, Atlanta, GA.

Table 7.1 Training in the Commissioning Process

Project Phase	Key Training Activities	Key Document Linkages
Predesign	Use the OPR development process to establish training needs.	Describe training needs in the Owner's Project Requirements.
	Develop a preliminary Training Plan.	Incorporate the first-cut Training Plan in the Commissioning Plan.
		Incorporate training activities to be conducted by design professionals in service agreements.
Design	Refine the understanding of training needs and develop a detailed training program.	Refine and adjust (as necessary) training requirements in the OPR.
	Incorporate training to be conducted by the contractor in the Construction Documents.	The Commissioning Plan is updated to include a final Training Plan.
	A detailed plan for training should be in place.	Construction Documents include all contractor-provided training requirements.
		Verification procedures for evaluating training effectiveness are established.

(continued)

Table 7.1 *(Continued)*

Project Phase	Key Training Activities	Key Document Linkages
Construction	The majority of training of owner's personnel is conducted.	The training aspects of the OPR are refined as required by design decisions.
	Training effectiveness is verified.	The Commissioning Plan is updated to include the most current Training Plan.
	Training is documented and materials are archived.	Details of training to occur during the occupancy and operations phase are added to the Training Plan (and thus to the Commissioning Plan).
		The bulk of the Systems Manual is completed and training information is integrated into the manual.
Occupancy and Operations	Conduct training scheduled for this phase—and/or deferred from the construction phase.	Update the training aspects of the OPR (as necessary).
	Train occupants/users as per the Training Plan.	Update the training aspects of the Commissioning Plan via the Training Plan (as necessary).
	Train new personnel.	Complete and use the Systems Manual.
	Retrain personnel as appropriate to add new skills or refresh existing skills.	

Chapter 8

Special Commissioning Contexts

SPECIAL CONTEXTS?

This chapter looks at three special contexts for building commissioning—ongoing commissioning, retrocommissioning, and commissioning of green projects. It also addresses discipline-specific commissioning resources that have been, or are being, developed to assist commissioning providers. Ongoing commissioning is essentially the extension of a "completed" commissioning process well into the life of a facility. Retrocommissioning addresses the commissioning of an existing facility that was not previously commissioned. The commissioning of green projects can be critical to the delivery of high-performance expectations, and is a prominent requirement of the U.S. Green Building Council's LEED-NC™ and LEED-EB™ certification programs (among others).

Although historically aimed at "active" systems (primarily HVAC-related, and to some extent electrical), building commissioning is currently seeing increasing acceptance as a process that can apply equally well to "passive" systems such as building envelope assemblies. Development of guidelines for the commissioning of systems beyond HVAC is briefly discussed.

ONGOING COMMISSIONING

Ongoing commissioning is a continuation of the commissioning process substantially into the life of a building—beyond the point of warranty expiration that often defines the termination of the basic commissioning process. The objective of ongoing commissioning is to extend the benefits derived from the commissioning process in order to maintain and/or improve (perhaps even to truly optimize) facility performance as a building is used over time. The verifications, benchmarking, and documentation provided by the conventional commissioning process are an excellent (and necessary) foundation for such an effort.

If commissioning is looked at as an investment in improved building performance (functionally, economically, and environmentally), rather than as just a way to get a contractor to produce an acceptable outcome, then the logic of continuing the commissioning process into the operating life of a project is inescapable. There is simply no serious rationale for terminating a successful process just because the design team and contractor are no longer involved. The intent of ongoing commissioning is the same as for conventional commissioning—to ensure that the Owner's Project Requirements are met. The focus of attention shifts, however, from verification of design decisions and equipment/system installations to systems operations and benchmarking.

As is the case during the design and construction phases, the Owner's Project Requirements may (in fact, are very likely to) morph or even radically change over the course of a building's life. Energy and environmental issues may become more important, economic forces may change, perceptions of acceptability for workplace conditions may evolve, or the basic function of large areas of a building may be altered. All such changes should be captured in an updated Owner's Project Requirements (OPR) document. The ability of installed and operating equipment and systems to meet such changes can be rationally addressed by comparison to this living OPR. Equally beneficially, on a day-to-day basis ongoing commissioning permits the performance of systems and components to be readily compared against the benchmarks established by the verification activities performed during the construction process. Equipment operation and system performance that are seen to be wandering away (usually in a negative direction) from these documented benchmarks can be investigated and corrected before problems escalate, substantial energy resources are wasted, or building environmental conditions noticeably degrade.

An owner with the capabilities to do in-house design/construction commissioning would also have the ability to do in-house ongoing commissioning. An owner without such capabilities would have engaged a commissioning authority to lead the commissioning process, and would presumably do the same for ongoing commissioning. Unless the experience was a serious disaster, it makes sense to engage the original commissioning authority to lead the ongoing commissioning process. Terms for such a contract would be negotiated between the parties to reach a mutually agreeable arrangement. A 1-year term for such an agreement is likely too short; a 10-year term likely too long. The level of anticipated effort (expectations and scope) would also be negotiated. Upon reaching agreement on the general scope and depth of ongoing commissioning, the development of an "Ongoing Commissioning Plan" is an appropriate first step in the process.

As with the conventional commissioning process, the commissioning authority will act as the leader of a commissioning team that will include owner's operating personnel, facility managers, users, and specialized consultants (as required). In opposition to the conventional commissioning process (which can require substantial time commitments at key points throughout the process), the time demands for ongoing commissioning should be moderate (except when serious changes in facility function and/or upgrading of equipment are planned).

Although there is no published guide aimed specifically at executing the ongoing commissioning process, logic and common sense should substitute as a reasonable guide. The commissioning authority would be expected to convene regularly scheduled (even if fairly widely spaced) meetings of the commissioning team (the composition of which will adapt to current circumstances). Minutes from such meetings should be taken, distributed, and incorporated into the already-existing project Commissioning Report. Running an "Ongoing Issues Log" is suggested as a useful documentation tool and communications vehicle. Updates to the Owner's Project Requirements, Basis of Design, and Training Plan should be made as needed. Updated test reports and new materials would be incorporated into the Systems Manual as necessary and appropriate. Training on revised approaches to operations/maintenance and or new equipment would also be conducted as required.

RETROCOMMISSIONING

ASHRAE Guideline 0 defines retrocommissioning as: "The commissioning process applied to an existing facility that was not previously commissioned" (ASHRAE 2005). Although defined by Guideline 0, no details for such an application of commissioning are given. Commissioning in this context will, however, surely require that a basis for verification be developed (a retro-OPR); that information about installed equipment, systems, and assemblies be collected, organized, and synthesized (a retro-Basis of Design); and that some level of detail regarding system and assembly performance be collected (a retrotesting regime)—so that the verification of observed on-site conditions against owner objectives that is the heart of commissioning can be accomplished. Retrocommissioning must go beyond simply determining whether something is working to establish whether it is working as it should. Figuring out what "as it should" actually means requires backfilling the missing parts of the never-completed original-project-acquisition commissioning process. The best formal guide for such an effort may be the *California Commissioning Guide: Existing Buildings* (California Commissioning Collaborative, 2006). Later in this chapter there is discussion of an ASHRAE guideline for existing HVAC&R systems that is now under development.

A desire to retrocommission may be triggered by a major change in building function or context, a serious or aggravating problem with operations, or a change in owner/operator philosophy (a desire to become more "green," for example). Or maybe an owner at some point just sees the light. Although retrocommissioning may be challenging and not the best way to reap the benefits of the commissioning process, it is never too late to do the right thing.

Recommissioning is a similar sounding, but conceptually very different, process. It is defined by ASHRAE as: "An application of the Commissioning Process requirements to a project that has been delivered using the commissioning process." Recommissioning should (unless all documentation has been lost) be able to build on the information assembled as part of the original commissioning process. It is a continuation of the commissioning process following some period of interruption. By contrast, ongoing commissioning is a continuation of the commissioning process—without interruption—following project turnover.

As with retrocommissioning, a decision to recommission may be triggered by a major change in building function or context (unassigned floor space becomes equipment-intensive production space), a problem with operations (energy costs are excessive), or a change in owner/operator philosophy (a desire to upgrade office quality to increase market competitiveness). Recommissioning may also simply reflect a delayed startup of the ongoing commissioning process. The key to recommissioning is that commissioning process artifacts (such as an Owner's Project Requirements, Basis of Design, System Manual, Construction Checklists) are available and can provide the jumping-off point for further commissioning efforts. The ability to tap into existing commissioning process documents makes recommissioning substantially easier than attempting to commission an existing building not previously commissioned (i.e., retrocommissioning—where much precursor work would need to be done simply to get information and documentation up to speed).

COMMISSIONING FOR GREEN BUILDINGS

The LEED green building rating systems, developed by the U.S. Green Building Council, have without a doubt been instrumental in bringing commissioning into the sights of many owners and design professionals who might otherwise not have considered commissioning a project. LEED for new construction (LEED-NC), LEED for commercial interiors (LEED-CI), and LEED for core and shell (LEED-CS) all require commissioning as a prerequisite for a green rating and provide an opportunity for a credit to be earned through enhanced commissioning. This is to be commended. The commissioning requirements in LEED-NC are generally typical of these rating systems (including the building-type-focused ratings for schools, retail, and healthcare) and will be discussed in some detail.

For new construction, addressed by LEED-NC, a project seeking any of the LEED certification levels must undergo "fundamental commissioning." Fundamental commissioning is described by the USGBC as having these minimum attributes:

- A commissioning authority will be engaged for the project. The authority may not be a member of the design team or construction team for the project, but may be employed by one of the design firms or the contractor (or a subcontractor) for the project. The independence of action of such an entity is somewhat unclear. Many believe this permits the engagement of a semi-independent party. If this is a concern, it is easily addressed by selecting a fully independent commissioning authority (which is advised by most commissioning practitioners). As with building codes, the requirements of LEED represent a minimum threshold that can be exceeded by an owner.
- The commissioning authority develops and implements a Commissioning Plan for the project. The activities described in this plan are reflected in subsequent efforts.
- Design intent and basis of design documentation (in ASHRAE Guideline 0 language, the Owner's Project Requirements and the Basis of Design) that are developed by the design team will be reviewed by the commissioning authority. Presumably, *reviewed* in this context means *verified*.
- The commissioning authority ensures (again, this suggests *verifies*) that appropriate commissioning requirements are included in the construction documents (which are prepared by the design team).
- The commissioning authority verifies equipment and systems installations, oversees *functional testing* (a throwback term not used in Guideline 0), coordinates training, and documents that the building meets design intent. In essence, these are the core activities of construction-phase commissioning.
- The commissioning authority develops and submits a commissioning report upon building occupancy (USGBC 2008a-d).

Fundamental commissioning as described and required by the LEED-NC prerequisite is not substantially at odds with the commissioning process as described in *ASHRAE Guideline 0*—with one key exception, a requirement to verify design-phase work products (other than the Basis of Design) is explicitly not included. The use of Guideline 0 (less design verification) as a roadmap for the LEED fundamental commissioning process is appropriate and strongly recommended. Differences in terminology (as noted above) seem trivial (as long as they are understood by the participants).

Beyond the fundamental commissioning prerequisite that underlies a LEED-NC certification, a credit may be obtained for undertaking "additional commissioning." The nature of this additional effort is as follows:

- The commissioning authority conducts a focus review (verification) of the design before the construction documents for the project are developed. A second review (verification) occurs near substantial completion of the construction documents. This effort brings the collective commissioning process (fundamental plus additional) into substantive conformance with the scope and depth of the commissioning process prescribed in *ASHRAE Guideline 0.*
- The commissioning authority reviews a selected sampling of equipment submittals. This activity enriches the construction-phase commissioning effort, and is implicit in the commissioning process defined by ASHRAE.
- The commissioning authority develops a manual to assist with recommissioning the project. In theory, the Systems Manual and final Commissioning Process Report described in Guideline 0 should fulfill this requirement.
- The commissioning authority conducts a post occupancy review of the building before equipment warranties expire (the specific period to be determined by the warranty terms). This activity is expressly required by Guideline 0.

A comment regarding commissioning for LEED projects is in order. A common refrain that seems to recur on such projects goes something like this: "What's the minimum that's really required?" The better question would be, "How can we use this prerequisite (and credit) to improve the building's performance?" A recent conference presentation on an interesting building made a compelling point in this regard. The wrong glazing was installed in the building in question, with serious repercussions for resulting performance. When asked why the commissioning process did not catch this error, the response was that envelope commissioning was not included because it was not required. Looking at the context and opportunity presented by commissioning will be much more fruitful than focusing on what's needed to get a credit. If a feature is critically important to building performance, it is critically important that it be commissioned—in any project, but most certainly in a green project.

LEED-EB (LEED for Existing Buildings) does not require commissioning as a prerequisite for certification. It does, however, offer up to 6 credits (of 30 possible) under the Energy and Atmosphere heading for "Existing Building Commissioning":

- Investigation and Analysis (2 credits)
- Implementation (2 credits)
- Ongoing Commissioning (2 credits)

Although the credits available from undertaking the commissioning process are substantial, as already noted, the greater benefit to all involved is likely to come from the inherent benefits of the commissioning process—better overall building performance. In most existing building situations, the commissioning process to be undertaken will be one of retrocommissioning (as reflected in the "Investigation and Analysis" credits). It is somewhat unclear why investigating without implementing would be considered creditworthy, but it is a start.

The Green Building Initiative, developers of the Green Globes environmental assessment system, issued a public review draft of its proposed *Green Building Assessment Protocol for Commercial Buildings* in April 2008. Building commissioning, *total* building commissioning in fact, as described in *ASHRAE Guideline 0* (and by extension, various discipline-specific guidelines) plays an important part in the protocol as published in draft form (GBI 2008).

ASHRAE Standard 189.1, *Standard for the Design of High-Performance, Green Buildings Except Low-Rise Residential Buildings*, is also being developed to provide a minimum set of requirements for a green building written in code-based language. The idea is that as a standard, these requirements can be easily adopted into code by an interested jurisdiction. This standard was out for a second public review as this is written. It appears that the final standard will require building commissioning (in general accordance with *ASHRAE Guideline 0*) for all projects deemed to comply with the standard.

Because backsliding on green building assessment and rating scheme requirements seems very unlikely, it is fair to say that the commissioning process will be a necessary element of "recognized" green projects for the foreseeable future. This will continue to be a key driver for commissioning services.

DISCIPLINE-SPECIFIC COMMISSIONING GUIDANCE

ASHRAE Guideline 0 was developed to describe the characteristics of an exemplary commissioning process, without regard to the specific element or system being addressed. It appears to have successfully assumed this role. Going beyond the generic process, however, there are details specific to a particular system (such as daylighting or elevators or roofing) that will assist the commissioning team in applying even a well-defined process to a particular situation. It is intended that these details be developed as discipline-specific guides (for lighting, plumbing, fire protection, landscaping, etc.) by appropriate professional societies and published via the NIBS Total Building Commissioning Guidelines series. This effort, a major undertaking both in terms of expertise and coordination, is well underway. Several components are discussed in this section.

ASHRAE has completed development of a revised version of Guideline 1 (its historic commissioning guideline, prior to the advent of Guideline 0). The purpose of this updated guideline is to focus on the HVAC systems that are at the core of ASHRAE's interests. The title of this guideline reflects this focus: *HVAC&R Technical Requirements for the Commissioning Process*. The purpose is to provide HVAC&R-specific guidance for the commissioning of such systems. Guideline 1.1 becomes the third in a series of total building commissioning guidance documents being developed under the auspices of the National Institute of Building Sciences. The first in series is *ASHRAE/NIBS Guideline 0*. The second (its numbering notwithstanding) is *NIBS Guideline 3*.

NIBS Guideline 3: Exterior Enclosure Technical Requirements For the Commissioning Process, was developed by NIBS to address the commissioning process as applied to the generally static elements of building enclosure. Published in 2006, it is available from NIBS via the Whole Building Design Guide Web site (NIBS 2006). As is true of ASHRAE Guideline 1.1-2007, and will very likely be the case with future discipline-specific guidelines, the bulk of Guideline 3 resides in the annexes that describe application details. As this guideline provides a good example of what to expect from forthcoming guidelines, it is appropriate to show its structure as reflected in the table of contents. The body of the guideline is roughly 36 pages of the 337-page document, so the annexes make up the bulk of the content.

Table of Contents for NIBS Guideline 3

Foreword
Purpose

Table of Contents for NIBS Guideline 3

(continued)

The Illuminating Engineering Society of North America (IESNA) recently launched a committee charged with developing a NIBS-based guideline for commissioning of lighting systems (electric and daylight). Unfortunately, a few months after start-up the committee seems to be on hold; time will tell how this topic progresses. The National Fire Protection

Association (NFPA) has been working on a NIBS-based guideline for commissioning of fire protection systems. Other discipline-specific guides are very likely to be developed under the auspices of NIBS, although it is hard to predict for which disciplines and when.

ASHRAE is currently developing a guideline that will address the commissioning process for existing HVAC systems. Originally numbered and titled *Guideline 30: The Commissioning Process for Existing HVAC&R Systems*, this was renumbered Guideline 1.2 to reflect its role in a coordinated series of commissioning guidelines. It is likely that this document will provide general guidance for the commissioning of existing systems (beyond the HVAC&R systems embedded in the title) and also serve as a good guide to retrocommissioning procedures (although it is not exclusively being written for retrocommissioning).

ASHRAE is also currently developing a guideline to provide assistance with the owner's personnel training that is an integral part of the commissioning process. Originally designated Guideline 31, this will be *Guideline 1.3: Building Operation and Maintenance Training for the HVAC&R Commissioning Process*.

The renumbering of ASHRAE guidelines is part of a long-term plan to synthesize ASHRAE guidelines dealing with commissioning into an organized and similarly formatted series. Thus, current Guideline 4 (addressing operations and maintenance documentation) may become Guideline 1.4 and (upon future revision) address the preparation of the Systems Manual. Existing Guideline 5 (*Commissioning Smoke Management Systems*) may become Guideline 1.5 and more closely link with the format of Guideline 0. From the level of activity described here, the future of building commissioning seems squarely on track for serious expansion and improved clarity.

INTERVIEW

Views on the Commissioning Process

H. J. Enck, LEED AP, CxAP
Principal and Founder of Commissioning & Green Building Solutions, Inc. (CxGBS), Atlanta, Georgia

Q. What is the general role of commissioning in the green building acquisition process?

A. The full commissioning process includes defining the end goal at the beginning of the project and involves continuous checks through the design, construction, and post-occupancy stages of a facility/project's life. Owners and their design, construction, operations and maintenance teams need the full commissioning process to efficiently and economically deliver and maintain the performance of a green/high performance building.

Teams embarking on a goal of delivering a green building need a clear vision of what they are being asked to deliver and what procedures are necessary for maintaining the performance of the building/project for its life. This starts with defining and documenting the owner's objectives, criteria, and end goals for the project, which make up the Owner's Project Requirements (OPR). This is a key document for success.

The OPR guides all design, construction, and operations decisions for a project, and helps save time and money. It includes the model that an owner will use to select green features that fit within the business model. This document puts the entire team on the same page from the very beginning—defining the goals of the project and establishing benchmarks for success, as well as developing design criteria to meet the needs of the building's occupants in fulfilling their daily mission. The team can then help owners select green features that fit within their business model. An example of an OPR can be downloaded from http://images.ashrae.biz/renovation/documents/opr61507final.pdf.

The commissioning process includes the following steps:

1. *Checking during the design phase to see that owners' project requirement goals are incorporated*
2. *Verifying during the construction phase that the design intent and requirements are correctly interpreted and implemented by contractors*
3. *Ensuring that building operators understand those features and how to maintain the performance of the building*
4. *Following up by monitoring and assisting owners with maintaining performance of the building over time.*

Q. Are there any fundamental differences between commissioning for a green building and a conventional building?

A. *In essence, there is no fundamental difference. The only difference is identifying sustainable goals that should really be part of every conventional building.*

Commissioning of the building envelope, mechanical, and electrical systems is important for a green building. The building envelope is key, because it has a significant impact on energy efficiency and indoor air quality. The interaction between the building envelope and HVAC system is critical for good indoor air quality.

Our CxGBS proprietary Holistic Commissioning™ Design Phase Commissioning Model deals with the building envelope, as well as traditional electrical, mechanical, and plumbing systems. This holistic approach lowers design, construction, and operational risks, improves the delivery process, and enhances occupants' physiological and psychological perceptions of their built environment.

Q. What seems to be the biggest misperception regarding commissioning for green buildings?

A. *I believe it is the cost of commissioning related to the return on investment. A commonly cited barrier to widespread use of commissioning is the decision maker's uncertainty about its cost-effectiveness.*

Building performance problems are pervasive and can lead to a host of costly ramifications, ranging from compromised indoor air quality with related health issues to unnecessarily elevated energy use and loss of property value. Solving problems during the design phase while designers and contractors are engaged saves an owner more than the cost of commissioning.

By making a decision to undertake design phase commissioning for its North and West Building projects in the Capital Complex, the State of Louisiana Division of Administration Facility Planning and Control (DAFPC) got a return on its investment of over 28 times.

The 28 times return on investment was calculated by dividing the cost of the design phase commissioning (approximately 0.12 percent of the total project budget of $100 million, or $0.12 per square foot) by the total commissioning-related savings (calculated as a result of findings), which totaled $3.4 million.

A 2004 study sponsored by the U.S. Department of Energy titled, "A Meta-Analysis of Energy & Non-Energy Impacts in Existing Buildings & New Construction in 2004" showed that median commissioning costs of $1.00/ft^2 (0.6 percent of total construction costs) for new construction yielded a median payback time of 4.8 years (excluding quantified nonenergy impacts).

For existing buildings, median commissioning costs of $0.27/ft^2 yielded whole-building energy savings of 15 percent and payback times of 0.7 years.

Another misperception is that commissioning can be done at the end, rather than at the beginning, of a project. To be fully effective and obtain maximum value, commissioning must begin in the predesign phase, where the value of starting with the end goal in mind can maximize team performance and reduce costs.

A further misconception is that commissioning is something that designers or contractors are already doing. Typically, designers' quality control practices leave a contractor with a number of questions, which typically mean higher costs. Contractors seeing a poor set of construction documents can bid low and enhance their profits with change orders capitalizing on an insufficient or poorly defined scope of work. Architects are typically focused on the visual elements of a project, and miss the really important needs of the users that are necessary for the building spaces to fulfill the mission of the occupants. These requirements include lighting function relative to space usage, temperature control limits within the space, acoustical requirements, and types of activities that will be conducted in conference rooms (such as audiovisual presentations, group meetings, video/tele-conferences, etc.).

Q. Are there any common errors you see regarding the application of commissioning to green buildings?

A. *Common errors I see are project teams not focused on fulfilling the Owner's Project Requirements for a facility. In some instances, project teams move forward using the same design practices they have used for years based on their own experiences and criteria instead of shifting their paradigm in order to meet what the owner has identified as the objectives and requirements for the project.*

The alignment of green goals with the owner's mission is typically missed. Project team members might advise owners of goals they believe should be accomplished to create a green building. Trusting these professionals with providing guidance regarding the right direction, the owner may go down the wrong path, pursuing costly goals that are not aligned with the specific purpose of the facility.

Other common errors include not beginning the commissioning process during the predesign phase, which creates lost opportunities, rework, and change orders that can ultimately increase costs for conventional projects and result in higher costs when green/sustainable goals are not met.

Q. What do you recommend as a means of breaking the ice on green building commissioning—for the novice owner?

A. *Helping owners review their business case objectives to develop a realistic budget for developing a high performance building—including building commissioning—is the best strategy for breaking the ice.*

When building green, owners need to first define what their business case is for a return on investment. For example, a retailer might consider daylighting if the building lease was for 10 years, as opposed to a retailer with a 5-year lease who needs a 3-year return on investment.

Owners should set aside a commissioning budget within their comfort level—generally estimated to be 1.0 to 1.5 percent of construction costs.

A definition of business case objectives and budgets in the Owner's Project Requirements guides the project team and helps them determine what green/sustainable strategies are best to apply to the owner's criteria. They can then communicate that information to the owner, who should make sure that the recommendations are in sync with the original goals for the facility.

Q. What resources regarding building commissioning do you recommend?

A. From a human resources standpoint, someone knowledgeable in commissioning the building envelope and mechanical and electrical systems should be part of the project team, beginning in the predesign phase.

Information resources may include reference guides and ASHRAE publications.

Q. What, if anything, new is on the horizon for green building commissioning efforts?

A. The trend is to focus on significantly improving the performance, operations, and maintenance of existing buildings.

The federal government has a mandate to reduce energy consumption by 30 percent over the next 10 years in GSA buildings.

Developers with larger commercial real estate holdings are starting to closely examine commissioning.

Q. Do you have any words of wisdom for those newly embarking on the green building commissioning process? Or for seasoned professionals?

A. Seasoned professionals, or those embarking on the green commissioning process for the first time, should start with their end goal in mind. They should select their sustainable goals based on a good business case for each project.

They should develop an Owner's Project Requirements document that is specifically aligned with these business goals.

It is critical to select a properly qualified commissioning authority with practical experience. This professional should be included on the team from the project's inception.

As the project progresses, it is important to ensure that good quality control processes are in place and that the sustainable development goals identified in the Owner's Project Requirements are being met.

REFERENCES

ASHRAE. 2005. *ASHRAE Guideline 0: The Commissioning Process*. The American Society of Heating, Refrigerating and Air-Conditioning Engineers, Atlanta, GA.

California Commissioning Collaborative. 2006. *California Commissioning Guide: Existing Buildings*. Sacramento, CA. www.documents.dgs.ca.gov/green/commissionguideexisting.pdf

GBI. 2008. Green Globes: draft of a proposed American National Standard. Green Building Initiative, Portland, OR. www.thegbi.org/home.asp.

NIBS. 2006. *Exterior Enclosure Technical Requirements For the Commissioning Process*. National Institute for Building Sciences, Washington, DC. www.wbdg.org/ccb/NIBS/nibs_gl3.pdf.

USGBC. 2008a. LEED Checklist—New Construction, LEED for New Construction (version 2.2). U.S. Green Building Council, Washington, DC. www.usgbc.org/DisplayPage.aspx?CMSPageID=220.

USGBC. 2008b. LEED Checklist—Existing Buildings, LEED for Existing Buildings (version 2.0). U.S. Green Building Council, Washington, DC. www.usgbc.org/DisplayPage.aspx?CMSPageID=220.

USGBC. 2008c. LEED Checklist—Commercial Interiors, LEED for Commercial Interiors (version 2.0). US Green Building Council, Washington, DC. www.usgbc.org/DisplayPage.aspx?CMSPageID=145.

USGBC. 2008d. LEED Checklist—Core & Shell, LEED for Core & Shell (version 2.0). U.S. Green Building Council, Washington, DC. www.usgbc.org/DisplayPage.aspx?CMSPageID=295.

Glossary

This book, as noted in the introductory chapters, is based on the commissioning process as described in *ASHRAE Guideline 0-2005*. Guideline 0 should be used as the official "dictionary" for formal definitions of commissioning process terms. Nevertheless, it is often good to get a second take on the basics—which is the intent of this glossary. The following explanations are not intended to duplicate the Guideline 0 definitions—or to contradict them

The term *formal document* as used here implies a written document, generally separate from other documents (although it may later be packaged with other documents) intended for review, verification, and acceptance as part of the commissioning process. *(Note:* Capitalized terms in the text are used to indicate a formal document or commissioning process deliverable.)

acceptance: The act of "approving" (undertaken by one with the contractual authority to do so) a report, document, or work activity. The critical implication of acceptance is that someone will get paid and/or work may proceed. The practical implication (ideally) is that someone has seen evidence of work at a level of completion and quality that is per their expectations and the scope of a contract. Thus, an owner will be expected to accept a draft Commissioning Plan, the Construction Documents, and the completed building. [Some accounts of commissioning that predate Guideline 0 describe an *acceptance phase* for a project. Such a phase, if accurately represented, would be project-long, as acceptance occurs from predesign through occupancy and operations.]

ASHRAE Guideline 0-2005: The Commissioning Process: This consensus guideline lays out the necessary components of the building commissioning process. It has generally been accepted as a reasonable statement of the ideal commissioning process by most—but not all—involved with commissioning in the United States. Guideline 0 is on continuous maintenance status, which means that revisions are considered upon receipt and can be published as necessary (instead of on a multi-year cycle). Owners should require that commissioning proposals from prospective providers substantially comply with Guideline 0, and that any exceptions be clearly identified. It is reasonable to expect some flexibility in process from project to project (and from provider to provider)—that is why "0" is a guideline and not a standard—but the intrinsic components of the process as defined by Guideline 0 need to be in place for a truly successful outcome.

Basis of Design: A formal document that describes why design solutions were developed as they were. The idea behind this document is to provide an easily accessible and transferable (versus random notes in a file drawer) record of critical design thinking. The Basis of Design is prepared by the design team, verified by the commissioning authority, should be accepted by the owner, is shared with the contractor (but not as a legally binding part of the construction documents), and is passed on to the owner's operating personnel for use during operation and maintenance of a facility. Examples of valuable information that would appear in a Basis of Design and in no other "shared" document include a statement of the codes used for design, the design conditions used, safety factors applied, assumptions about maintenance and replacement access, and the like.

checklist: A preprepared template that provides a script by which a design or construction activity is verified during the commissioning process. The idea behind checklists is to ensure that the verification process is well-structured in advance. This is not to preclude extemporaneous observations or concerns, but rather to ensure that critical issues are not overlooked in ad lib reviews. Checklists also provide a consistent and easily filed/retrieved record of verification efforts. *ASHRAE Guideline 0* focuses primarily upon Construction Checklists (see below), but there is no reason why Design Checklists could/should not also be used.

commissioning authority: The name given to the lead contact for the commissioning process. Rarely (except for small projects) is the commissioning process undertaken by a lone individual. It is important, however, to have a point of contact through which all information and decisions related to the commissioning firm and commissioning team flow—this is the commissioning authority. The commissioning authority on most projects will be responsible for the actions of several professionals from a commissioning firm, as well as being the leader of the commissioning team (see below). [Some accounts of commissioning that predate Guideline 0 use the term commissioning agent to describe this role. This term is discouraged, as it implies that this party has legal permission to act in lieu of the owner, which is typically not the case. An example: the commissioning authority acts on behalf of an owner, but seeks the owner's acceptance of work; a commissioning agent could simply accept work as if he/she were the owner.] The term commissioning professional is also encountered in practice.

Commissioning Plan: A formal document, which at any time during the commissioning process describes the broad picture of the commissioning process for a project, shows in detail the commissioning activities for the coming project phase (with schedule, budget, and proposed roles), and summarizes activities completed during previous phases. The Commissioning Plan is the roadmap for the commissioning process on a given project. The map is general for more-distant phases, and is detailed for the next-upcoming phase. The Commissioning Plan—which evolves throughout the life of a project—is prepared by the commissioning authority, accepted by the owner, and shared with all parties to the process (owner, design team, contractor, specialists, owner's operating personnel, etc.). The Commissioning Plan is an important informational tool—but it is not a contractually binding requirement upon those parties it mentions. Professional service agreements and construction contracts must require participation in the commissioning process (and should refer to the Commissioning Plan either by addendum or as background information).

Commissioning Report: This term describes an ongoing series of reports that document the commissioning process for a project. During the course of a project, Commissioning Progress Reports are prepared and distributed to interested parties. At the end of a project, a Final Commissioning Process Report is prepared that summarizes the complete commissioning process.

commissioning team: A loose-knit, but critically important, coalition of people involved with a project—formally assembled to complete and/or review commissioning process activities. The size and composition of the commissioning team varies throughout the course of a project (in response to evolving project needs). Skilled use of the commissioning team approach to commissioning can change the perception of commissioning from "me versus you" to one of "us versus unacceptable outcomes." It sounds a bit corny, but teamwork will spread buy-in and also improve decision making through diversity of inputs. The commissioning team functions primarily through the venue of commissioning team meetings—called by the commissioning authority (who should set the agenda) and staffed by the various parties to the activity at hand (in response to requests from the commissioning authority). Good planning and prescheduling of meetings will facilitate active involvement. Meeting minutes should be taken and distributed and action items tracked for resolution.

Construction Checklists: These are checklists specifically developed for use in verifying the proper installation and operations of materials, equipment, and systems during the construction phase. See *Checklists* for more information. Checklists are conceptually related to OPR-focused *test procedures*.

Construction Documents: Normally, these consist of the project drawings and specifications, which convey information on what is to be built. This information is conveyed from the design team to the contractor via the owner. The drawings usually show the contractor what (generically) goes where. From this, the contractor establishes quantities of materials and equipment to be purchased. The specifications describe the quality of materials and their installation. Commissioning process requirements will be conveyed in the specifications—making this document critical to successful commissioning. Typically, neither the design team nor the owner can or should tell the contractor how to construct a building. The focus of construction documents is on outcomes, not process. Commissioning verifications cross this line a bit (a contractor can't be allowed to verify proper operation of a chiller just by listening to it); thus, coordination between the contractor and commissioning authority relative to verifications is essential.

contract documents: All the documents that contractually bind a contractor to complete a building project for an owner. Project drawings and specifications are a major element of the contract documents, but the contract documents may also include other legally binding requirements, such as a schedule, special requirements, and the like. Any owner desire, such as commissioning, that is not part of the contract documents will not happen—except as a contract addendum (which are typically very expensive).

contractor: The entity hired by an owner to build a project. For virtually all projects, the contractor will hire subcontractors to do substantial portions of the work. Suppliers will be engaged by the contractor to supply materials and equipment. All work to be done by the contractor must be explicitly called out in the contract documents. There are numerous ways that this call-out can be made, the most common being through drawings and specifications prepared by a design team hired by the owner.

construction drawings: A key component of the Construction Documents. Although commissioning process requirements will not appear in the drawings, quality drawings are essential to a high-performance, successful project. These drawings should be reviewed and verified against the Owner's Project Requirements during the design phase. A statistically based sampling method is recommended for this activity.

construction specifications: A key component of the Construction Documents. Commissioning process requirements related to the contractor will appear in the specifications, and these must be presented clearly and adequately to ensure appropriate contractor participation. In addition, quality specifications are essential to a high-performance, successful project. The specifications should be reviewed and verified against the Owner's Project Requirements during the design phase. A statistically based sampling method is recommended for this activity. Careful review of commissioning process requirements is recommended.

Design Checklists: See *Checklists*. Design Checklists are prepared by the commissioning authority to assist with verifications that occur during the design phase of a project.

drawings: See *Construction Drawings*. Coordination drawings can be especially useful to the commissioning process by showing the work of all disciplines—as a means of ensuring coordination between trades and the appropriate fit of equipment into allotted spaces (with due consideration to access).

Issues Log: A formal system for registering concerns raised during the commissioning process—and the resolution of such concerns. The use of an electronic database that can be easily queried is highly recommended. The Issues Log is a valuable part of the communications infrastructure that is a hallmark of successful commissioning.

ongoing commissioning process: The continuation of the commissioning process into the life of a project—beyond the "normal" termination point of warranty expiration. The idea behind ongoing commissioning is to continue to reap the benefits of the process during the extensive occupancy and operations phase of a project.

Owner's Project Requirements (OPR):   A formal document that describes in detail what is necessary for a project to be considered a success from the owner's perspective. This document goes beyond a typical spatial program to define the quality of spaces and their environments and operations. The OPR should address thermal, lighting, and acoustical conditions, reliability and maintenance expectations, energy benchmarks, green design expectations, and the like. Most owners don't just want 10,000 square feet of space—they want space with specific characteristics, which the OPR will describe. The OPR may be developed by the owner or by the commissioning team (depending on owner capabilities), will be verified by the commissioning authority, accepted by the owner, and given to the design team. The OPR should ideally be given to the contractor as "information only" to allow more informed decision making and be given to the owner's operating personnel for the same reason.

recommissioning: The application of the commissioning process to a building that was previously commissioned. Although generally similar in nature, recommissioning may be differentiated from ongoing commissioning by virtue of a "break" in the process.

retrocommissioning process: The application of the commissioning process to a building not previously commissioned. This scenario will require that the commissioning process develop all the key commissioning documents that would normally have been produced in the course of a conventional (starting at predesign) commissioning process. Retrocommissioning is not just a building tune-up exercise—it must identify project intents and criteria (OPR) against which current performance can be benchmarked.

sampling: A means of selecting limited areas of work for verification. Sampling is intended to reduce the volume of verification effort—without sacrificing the benefits of verification. The philosophy behind sampling is that the design team and contractor are legally responsible for their own quality control efforts and outcomes, but that another set of independent eyes looking at randomly selected facets of a project can help ensure overall quality.

specifications: See *construction specifications*.

Systems Manual: A formal (and well-prepared and organized) document intended to assist the owner in getting optimum benefit from his/her facility. In essence, this document is a good owner's manual for the facility. The responsibility for preparation of the Systems Manual must be established and expectations for the manual clearly communicated to the responsible party. Historically, this has been the contractor. Historically, this has not always led to the best product possible—although the contractor has ready access to most of the elements that will comprise the manual. A tight specification for this product is probably the best way to ensure a good product. A Systems Manual may be viewed as an enhanced O&M (Operations and Maintenance) Manual—including virtually anything the owner's operating personnel may need to know to get the best out of a facility (including training materials and performance benchmarking checklists and test results).

test procedures: Procedures developed in support of systems and assembly verification activities. Test procedures are used for more complex evaluations of performance than can be reasonably conveyed via a checklist. For example, a test procedure might be developed to ensure that the performance of a lighting system (including controls) meets the Owner's Project Requirements for a space (perhaps including solutions such as zoning, presets, daylight dimming, occupancy sensors, and/or lumen maintenance). Test procedures will typically be developed by the commissioning authority in cooperation with the contractor. Test records will provide a benchmark for ongoing system operations.

Training Plan: A subset of the Commissioning Plan that describes, in detail, the owner's personnel training requirements and expectations (including verifications) for a project. The Training Plan will describe scope, activities, schedule, roles, and evaluations involved with training personnel to effectively operate the facility they will soon be responsible for. The Training Plan will be developed by the commissioning authority (with input from the commissioning team) and approved by the owner.

Elements of the plan to be conducted by the design team and contractor must be included in professional services contracts and the construction documents.

verification: This term is used to describe the process whereby aspects of design and construction are evaluated for conformance with the Owner's Project Requirements. For example, can a proposed HVAC system deliver the indoor air quality, flexibility, comfort, and energy performance demanded by an owner? Is equipment installed in such a way that it can support such performance? Verification is a key responsibility of the commissioning authority (with assistance from the commissioning team).

Commissioning Resources

FUNDAMENTAL COMMISSIONING RESOURCES

ACG: Associated Air Balance Commissioning Group. Provides a certification exam testing basic knowledge and technical expertise of eligible commissioning providers and publishes the *ACG Commissioning Guideline.* www.aabchq.com/commissioning/

ASHRAE: American Society of Heating, Refrigerating and Air-Conditioning Engineers, Inc. Publisher of several commissioning guidelines—available through the online bookstore. www.ashrae.org

BCA. Building Commissioning Association. A professional organization that promotes commissioning and best practices therein. Some resources including publications and process templates (such as Construction Checklists and RFQs) are publicly available via its Web site. www.bcxa.org/resources/pubs/index.htm

Building Commissioning. A somewhat eclectic Web site with "articles, links and helpful info." http://buildingcommissioning.wordpress.com/

CACX. California Commissioning Collaborative. A nonprofit board that promotes commissioning in California and provides commissioning resources via an extensive Web site. www.cacx.org/

National Clearing House for Educational Facilities. School Building Commissioning. Provides a range of commissioning resources with a focus upon school facilities. www.edfacilities.org/rl/commissioning.cfm

NEBB: National Environmental Balancing Bureau. Provides certification of firms and qualification of individuals through its Building Systems Commissioning Program (addressing HVAC, plumbing, and fire protection systems). www.nebb.org/

Oregon Department of Energy—Conservation Division. Commissioning for Better Buildings in Oregon. Provides links to general commissioning resources. http://oregon.gov/ENERGY/CONS/BUS/comm/bldgcx.shtml

PECI. Portland Energy Conservation, Inc. Sponsors the annual National Conference on Building Commissioning and provides substantial resources for building commissioning. Look under "Commissioning and Technical Resources." www.peci.org/

University of Wisconsin-Madison, Department of Engineering Professional Development. A long-term provider of short courses on various aspects of commissioning; offers a certificate/certification program for commissioning providers. http://epdweb.engr.wisc.edu/

GENERAL COMMISSIONING ISSUES

AABC: *ACG Commissioning Guideline*, 2nd ed. Associated Air Balance Council, Washington, DC.

ASHRAE. 2008. ASHRAE Guideline 4-2008: *Preparation of Operating and Maintenance Documentation for Building Systems*. American Society of Heating, Refrigerating and Air-Conditioning Engineers, Inc., Atlanta, GA.

ASHRAE. 2001. ASHRAE Guideline 5-1994: *Commissioning Smoke Management Systems*. American Society of Heating, Refrigerating and Air-Conditioning Engineers, Inc., Atlanta, GA. [Reaffirmed in 2001.]

ASHRAE. 2005. ASHRAE Guideline 0-2005: *The Commissioning Process*. American Society of Heating, Refrigerating and Air-Conditioning Engineers, Inc., Atlanta, GA.

ASHRAE. 2007. *ASHRAE Handbook—2007 HVAC Applications*. Chapter 42 "HVAC Commissioning." American Society of Heating, Refrigerating and Air-Conditioning Engineers, Inc., Atlanta, GA.

ASHRAE. 2007. ASHRAE Guideline 1.1-2007: *HVAC&R Technical Requirements for the Commissioning Process*. American Society of Heating, Refrigerating and Air-Conditioning Engineers, Inc., Atlanta, GA.

ASHRAE. 20xx. Guideline 1.2-20xx: *The Commissioning Process for Existing HVAC&R Systems*. American Society of Heating, Refrigerating and Air-Conditioning Engineers, Inc., Atlanta, GA. [Under development; addresses retrocommissioning of HVAC&R systems.]

ASHRAE. 20xx. ASHRAE Guideline 1.3-20xx: *Building Operation and Maintenance Training for the HVAC&R Commissioning Process*. American Society of Heating, Refrigerating and Air-Conditioning Engineers, Inc., Atlanta, GA. [Under development; addresses the training aspects of commissioning with a focus on HVAC&R systems.]

Bochat, James. 2005. "Commissioning Communications." *Proceedings of the 2005 National Conference on Building Commissioning* (see entry for PECI). www.peci.org/ncbc/proceedings/2005/21_Bochat_NCBC2005.pdf

Brooks, Bradley. 2007. "No Operator Left Behind: Effective Methods of Training Building Operators." *Proceedings of the 2007 National Conference on Building Commissioning* (see entry for PECI). www.peci.org/ncbc/proceedings/2007/Brooks_NCBC2007.pdf

CACX: 2006. *California Commissioning Guide: New Buildings*. California Commissioning Collaborative, Sacramento, CA.

CACX: 2006. *California Commissioning Guide: Existing Buildings*. California Commissioning Collaborative, Sacramento, CA.

CIBSE. 2007. CIBSE Commissioning Codes (includes: CIBSE Commissioning Code A: Air Distribution Systems; CIBSE Commissioning Code B: Boilers; CIBSE Commissioning Code C: Automatic Controls; CIBSE Commissioning Code L: Lighting; CIBSE Commissioning Code M: Management; CIBSE Commissioning Code R: Refrigeration; and CIBSE Commissioning Code W: Water Distribution Systems). The Chartered Institution of Building Services Engineers, London, UK.

DOE. Building Technologies Program—Buildings Toolbox—Building Commissioning. U.S. Department of Energy, Washington, DC. www.eere .energy.gov/buildings/info/operate/buildingcommissioning.html

Elvin, George. 2007. *Integrated Practice in Architecture*. John Wiley & Sons, Hoboken, NJ.

Heinz, John, and Richard Casault. 2004. *The Building Commissioning Handbook*, 2nd ed. The Association of Higher Education Facilities Officers, Alexandria, VA.

Kettler, Gerald. 2005. "Coordinating Building Systems for Commissioning." *Proceedings of the 2005 National Conference on Building Commissioning* (see entry for PECI). www.peci.org/ncbc/proceedings/2005/ 21_Kettler_NCBC2005.pdf

Mauro, Frank. 2005. "Commissioning Basics for Owners." *Proceedings of the 2005 National Conference on Building Commissioning* (see entry for PECI). www.peci.org/ncbc/proceedings/2005/BF_01_Mauro_NCBC2005.pdf

NEBB. 1999. *Procedural Standards for Building Systems Commissioning*. National Environmental Balancing Bureau, Gaithersburg, MD.

NEBB. 2005. *Design Phase Commissioning Handbook*. National Environmental Balancing Bureau, Gaithersburg, MD.

NIBS. Whole Building Design Guide. "Project Management—Building Commissioning." National Institute of Building Sciences, Washington, DC. www.wbdg.org/project/buildingcomm.php

Oberlander, George. 2007. "The Nuts and Bolts of the Commissioning Process." *Proceedings of the 2007 National Conference on Building Commissioning* (see entry for PECI). www.peci.org/ncbc/proceedings/ 2007/Oberlander_NCBC2007.pdf

PECI. National Conference on Building Commissioning (papers from the conference proceedings from 2005 onward). Portland Energy Conservation, Inc., Portland, OR. www.peci.org/ncbc/proceedings.html

Rochwarg, Leah. 2005. "Legal Implications of Building Commissioning" (PowerPoint presentation). *Proceedings of the 2005 National Conference on Building Commissioning* (see entry for PECI). www.peci.org/ncbc/ proceedings/2005/07_Rochwarg_NCBC2005.pdf

SMACNA. 1994. *HVAC Systems—Commissioning Manual*. Sheet Metal and Air Conditioning Contractors' National Association, Chantilly, VA.

Stum, Karl, and Kent Barber. 2005. "Critical Commissioning Communication." *Proceedings of the 2005 National Conference on Building Commissioning* (see entry for PECI). www.peci.org/ncbc/proceedings/2005/21_Stum_NCBC2005.pdf

COSTS AND BENEFITS OF COMMISSIONING

Della Barba, Michael. 2005. "The Dollar Value of Commissioning." *Proceedings of the 2005 National Conference on Building Commissioning* (see entry for PECI). www.peci.org/ncbc/proceedings/2005/19_DellaBarba_NCBC2005.pdf

Ellis, Rebecca, et al. 2006. "Quantifying Costs and Benefits, the Sequel: A 6-Year Check-up on Commissioning at the Pentagon." *Proceedings of the 2006 National Conference on Building Commissioning* (see entry for PECI). www.peci.org/ncbc/proceedings/2006/06_Ellis_NCBC2006.pdf

Jewell, Mark. 2005. "Understanding the Value of Commissioning in Income Properties" (Power Point presentation). *Proceedings of the 2005 National Conference on Building Commissioning* (see entry for PECI). www.peci.org/ncbc/proceedings/2005/03_Jewell_NCBC2005.pdf

R.S. Means. 2006. *Green Building: Project Planning and Estimating*, 2nd ed. R.S. Means Company, Kingston, MA.

Mills, Evan, et al. 2005. Costs and Benefits of Commissioning New and Existing Commercial Buildings. PowerPoint presentation with extensive data. http://hightech.lbl.gov/Presentations/Mills_Cx_UCSC.ppt

Mills, Evan, et al. 2005. "The Cost-Effectiveness of Commissioning New and Existing Commercial Buildings: Lessons from 224 Buildings." Proceedings of the 2005 National Conference on Building Commissioning (see entry for PECI). www.peci.org/ncbc/proceedings/2005/19_Piette_NCBC2005.pdf

PECI. 2002. "Establishing Commissioning Costs." www.peci.org/Library/PECI_NewConCx1_1002.pdf

COMMISSIONING FOR GREEN BUILDINGS

Altwies, Joy. 2002. "Commissioning for LEED." http://resources.cacx.org/library/holdings/009.pdf

Berning, Michael, and Bart Grunenwald. 2005. "How to Achieve LEED Certification for Commissioning Projects." *Proceedings of the 2005 National Conference on Building Commissioning* (see entry for PECI). www.peci.org/ncbc/proceedings/2005/24_Berning_NCBC2005.pdf

Coto, Jorge. 2007. "What Needs to be Commissioned in Zero Energy Buildings?" www.bcxa.org/events/expo2007/bca-expo-2007-torrescoto.pdf

D'Antonio, Peter. 2007. "Costs and Benefits of Commissioning LEED-NC Green Buildings." *Proceedings of the 2007 National Conference on Building Commissioning* (see entry for PECI). www.peci.org/ncbc/proceedings/2007/ DAntonio_NCBC2007.pdf

FMLink. Facilities Management Resources—Sustainability—Green Building Commissioning. www.fmlink.com/ProfResources/Sustainability/ Articles/article.cgi?USGBC:200309-01.htm

GBI. 2008. Green Globes: Draft of a Proposed American National Standard. Green Building Initiative, Portland, OR. www.thegbi.org/home.asp

Mantai, Michael. 2006. "Expanded Role of the Commissioning Provider for LEED Projects." *Proceedings of the 2006 National Conference on Building Commissioning* (see entry for PECI). www.peci.org/ncbc/proceedings/ 2006/04_Mantai_NCBC2006.pdf

McCown, Paul. 2007. "Commissioning for LEED" (PowerPoint presentation). www.bcxa.org/events/expo2007/bca-expo-2007-mccown.pdf

PECI. 2002. "Commissioning of Smaller Green Buildings—Expectations versus Reality." www.peci.org/Library/PECI_SmallGreenCx1_1002.pdf

PECI. 2000. "Commissioning to Meet Green Expectations." www.peci.org/ Library/PECI_CxGreen1_0402.pdf

Sinopoli, Jim. 2007. "How Do Smart Buildings Make a Building Green?" AutomatedBuildings.com   www.automatedbuildings.com/news/dec07/articles/ sinopoli/071129114606sinopoli.htm

USGBC. 2008. LEED Checklist—New Construction, LEED for New Construction (version 2.2). U.S. Green Building Council, Washington, DC. www.usgbc.org/DisplayPage.aspx?CMSPageID=220

USGBC. 2008. LEED Checklist—Existing Buildings, LEED for Existing Buildings (version 2.0). U.S. Green Building Council, Washington, DC. www.usgbc.org/DisplayPage.aspx?CMSPageID=220

USGBC. 2008. LEED Checklist—Commercial Interiors, LEED for Commercial Interiors (version 2.0). U.S. Green Building Council, Washington, DC. www.usgbc.org/DisplayPage.aspx?CMSPageID=145

USGBC. 2008. LEED Checklist—Core & Shell, LEED for Core & Shell (version 2.0). U.S. Green Building Council, Washington, DC. www.usgbc .org/DisplayPage.aspx?CMSPageID=295

Wilkinson, Ronald. 2008. "LEED Commissioning for New and Existing Buildings." *HPAC Engineering* (online). http://hpac.com/mag/leed _commissioning_new/

RETROCOMMISSIONING

ACEEE. 2003. "Retrocommissioning: Program Strategies to Capture Energy Savings in Existing Buildings." American Council for an Energy-Efficient Economy, Washington, DC. www.aceee.org/pubs/a035full.pdf

CACX. Retrocommissioning Toolkit. California Commissioning Collaborative, Sacramento, CA. www.cacx.org/resources/rcxtools/index.html

LBNL. 2004. "The Cost-Effectiveness of Commercial-Buildings Commissioning: A Meta-Analysis of Energy and Non-Energy Impacts in Existing Buildings and New Construction in the United States." Lawrence Berkeley National Laboratory, Berkeley, CA. http://eetd.lbl.gov/emills/PUBS/Cx-Costs-Benefits.html

PECI. 1999. "A Practical Guide for Commissioning Existing Buildings." Portland Energy Conservation, Inc., Portland, OR. http://eber.ed.ornl.gov/commercialproducts/retrocx.htm

PECI. 2007. "A Retrocommissioning Guide for Building Owners." Portland Energy Conservation, Inc., Portland, OR. www.peci.org/Library/EPAguide.pdf

PECI. 2001. "Retrocommissioning's Greatest Hits." Portland Energy Conservation, Inc., Portland, OR. www.peci.org/Library/PECI_RCxHits1_1002.pdf

Peterson, Janice, Bendy Ho, and Hyoin Lai. 2005. "Evaluation of Retrocommissioning Results After Four Years: A Case Study." *Proceedings of the 2005 National Conference on Building Commissioning* (see entry for PECI). www.peci.org/ncbc/proceedings/2005/14_Peterson_NCBC2005.pdf

Poeling, Tom. 2006. "Tuning Up the Retrocommissioning Process." *Proceedings of the 2006 National Conference on Building Commissioning* (see entry for PECI). www.peci.org/ncbc/proceedings/2006/21_Poeling_NCBC2006.pdf

DISCIPLINE SPECIFIC COMMISSIONING

Beaty, Don. 2005. "Commissioning Raised Floors for UFAD Applications." *Proceedings of the 2005 National Conference on Building Commissioning* (see entry for PECI). www.peci.org/ncbc/proceedings/2005/TM_01_Beaty_NCBC2005.pdf

Darragh, Shaun. 2003. "Lighting Commissioning." From the Lighting Design Lab (Seattle). www.lightingdesignlab.com/ldlnews/lighting_commissioning_sd.pdf

Deringer, Joseph, et al. 2007. "Guideline 3-1006: Exterior Enclosure Technical Requirements for the Commissioning Process." *Proceedings of the 2007 National Conference on Building Commissioning* (see entry for PECI). www.peci.org/ncbc/proceedings/2007/Deringer1_NCBC2007.pdf

Dorgan, Charles. 2007. "Guideline 1-2007: HVAC&R Technical Requirements for the Commissioning Process." *Proceedings of the 2007 National Conference on Building Commissioning* (see entry for PECI). www.peci.org/ncbc/proceedings/2007/Dorgan_NCBC2007.pdf

Larsen, Trina. 2007. "Commissioning Advanced Lighting Systems." *Proceedings of the 2007 National Conference on Building Commissioning* (see entry for PECI). www.peci.org/ncbc/proceedings/2007/Larsen_NCBC2007.pdf

LBNL. 2007. "Daylighting the New York Times Headquarters Building (Final Report)." Lawrence Berkeley National Laboratory, Berkeley, CA. http://windows.lbl.gov/comm_perf/pdf/daylighting-nyt-final-III.pdf

Miller, David. 2007. "Commissioning Security Systems" (PowerPoint presentation). www.bcxa.org/events/expo2007/bca-expo-2007-miller.pdf

NIBS. 2006. *NIBS Guideline 3: Exterior Enclosure Technical Requirements for the Commissioning Process.* The National Institute of Building Sciences, Washington, DC. www.wbdg.org/ccb/NIBS/nibs_gl3.pdf

Phelan, John. 2002. "Commissioning Lighting Control Systems for Daylighting Applications." http://resources.cacx.org/library/holdings/178.pdf

Seattle, City of. "Standard Commissioning Procedure for Daylighting Controls." www.seattle.gov/light/conserve/business/bdgcoma/bca8.pdf

Seth, Anand, Abbe Bjorklund, and Daryl Fournier. 2006. "Commissioning Scope of Work for Critical Healthcare Facilities." *Proceedings of the 2006 National Conference on Building Commissioning* (see entry for PECI). www.peci.org/ncbc/proceedings/2006/14_Seth_NCBC2006.pdf

Steward, Shannon, Mark Leafstedt, and Mike Abrams. 2005. "Fire Life Safety Commissioning." *Proceedings of the 2005 National Conference on Building Commissioning* (see entry for PECI). www.peci.org/ncbc/proceedings/2005/20_Steward_NCBC2005.pdf

Tseng, Paul. 2005. "Commissioning the Windows: Design Phase Strategies for High Performance Buildings." *Proceedings of the 2005 National Conference on Building Commissioning* (see entry for PECI). www.peci.org/ncbc/proceedings/2005/12_Tseng_NCBC2005.pdf

Turner, William, et al. 2005. "Retro-Commissioning & Commissioning Building Envelope Systems to Reduce Health Risks & Improve IAQ: What We Have Learned To Date." *Proceedings of the 2005 National Conference on Building Commissioning* (see entry for PECI). www.peci.org/ncbc/proceedings/2005/16_Turner_NCBC2005.pdf

Index